THE PKU PARADOX

JOHNS HOPKINS BIOGRAPHIES OF DISEASE
Charles E. Rosenberg, Series Editor

Randall M. Packard, *The Making of a Tropical Disease: A Short History of Malaria*

Steven J. Peitzman, *Dropsy, Dialysis, Transplant: A Short History of Failing Kidneys*

David Healy, *Mania: A Short History of Bipolar Disorder*

Susan D. Jones, *Death in a Small Package: A Short History of Anthrax*

Allan V. Horwitz, *Anxiety: A Short History*

Diane B. Paul and Jeffrey P. Brosco, *The PKU Paradox: A Short History of a Genetic Disease*

The PKU Paradox

❖ ❖ ❖

A Short History of a Genetic Disease

Diane B. Paul and
Jeffrey P. Brosco

Johns Hopkins University Press
Baltimore

© 2013 Johns Hopkins University Press
All rights reserved. Published 2013
Printed in the United States of America on acid-free paper
2 4 6 8 9 7 5 3 1

Johns Hopkins University Press
2715 North Charles Street
Baltimore, Maryland 21218-4363
www.press.jhu.edu

Library of Congress Cataloging-in-Publication Data
Paul, Diane B., 1946–
The PKU paradox : a short history of a genetic disease / Diane B. Paul and Jeffrey P. Brosco.
   p. ; cm. — (Johns Hopkins biographies of disease)
Includes bibliographical references and index.
ISBN 978-1-4214-1131-6 (pbk. : alk. paper) — ISBN 1-4214-1131-8 (pbk. : alk. paper) —
ISBN 978-1-4214-1132-3 (electronic) — ISBN 1-4214-1132-6 (electronic)
I. Brosco, Jeffrey P. (Jeffrey Paul) II. Title. III. Series: Johns Hopkins biographies of disease.
 [DNLM: 1. Phenylketonurias—history. WD 200]
RC632.H87
616.3'99—dc23   2013010207

A catalog record for this book is available from the British Library.

Special discounts are available for bulk purchases of this book. For more information, please contact Special Sales at 410-516-6936 or specialsales@press.jhu.edu.

Johns Hopkins University Press uses environmentally friendly book materials, including recycled text paper that is composed of at least 30 percent post-consumer waste, whenever possible.

CONTENTS

Foreword, by Charles E. Rosenberg   vii
Preface   xiii
List of Abbreviations   xxiii

Introduction. Pearl Buck, PKU, and Mental Retardation   1

Chapter 1. The Discovery of PKU as a Metabolic Disorder   10

Chapter 2. PKU as a Form of Cognitive Impairment   22

Chapter 3. Testing and Treating Newborns, 1950–1962   35

Chapter 4. The Campaign for Mandatory Testing   54

Chapter 5. Sources of Skepticism   72

Chapter 6. New Paradigms for PKU   92

Chapter 7. Living with PKU   111

Chapter 8. The Perplexing Problem of Maternal PKU   140

Chapter 9. Who Should Procreate? Perspectives on Reproductive Choice and Responsibility in Postwar America   156

Chapter 10. Newborn Screening Expands   179

Epilogue. "The Government Has Your Baby's DNA": Contesting the Storage and Secondary Use of Residual Dried Blood Spots 204

Acknowledgments 213
A Note on Sources 219
Notes 223
Index 283

FOREWORD

Disease is a fundamental aspect of the human condition. Ancient bones tell us that pathological processes are older than humankind's written records, and sickness and death still confound us. We have not banished pain, disability, or the fear of death, even if we die, on the average, at older ages, of chronic and not acute ills, in hospital or hospice beds and not in our own homes. Disease is something men and women feel. It is experienced in our bodies—but also in our minds and emotions. It can bring pain and incapacity and hinder us at work and in meeting family responsibilities. Disease demands explanation; we think about it and we think with it. Why have I become ill? And why now? How is my body different in sickness from its unobtrusive functioning in health? Why in times of epidemic has a community been scourged? Why do some infants fail to thrive?

Answers to such timeless questions necessarily mirror and incorporate time- and place-specific ideas, social assumptions, and technological options. In this sense, disease has always been a social and linguistic, a cultural as well as biological entity. In the Hippocratic era more than two thousand years ago, physicians were limited to a sufferer's words and to the evidence of their own senses in diagnosing a fever, an abnormal discharge, or seizures. Their notions of the material basis for such felt and visible symptoms necessarily reflected and incorporated contemporary philosophical and physiological concepts, a speculative world of disordered humors, "breath," and pathogenic local environments. Today we can call for understanding upon a rather different variety of scientific understandings and an armory of diagnostic tools that allow us to diagnose ailments (and even the likelihood of ailments) unfelt by patients and imperceptible to the doctor's unaided senses. In the past century, disease has become increasingly a bureaucratic phenomenon as well—as sickness has been defined and in that sense constituted by formal disease classifications, treatment protocols, and laboratory thresholds.

Sickness is also linked to climatic and geographic factors. How and where we live and how we distribute our resources all contribute to the incidence of disease. For example, ailments such as typhus fever, plague, malaria, dengue, and yellow fever reflect specific environments that we have shared with our insect contemporaries. But humankind's physical circumstances are determined in part by culture—and especially agricultural practice in the millennia before the growth of cities and industry. Environment, demography, economic circumstances, and applied medical knowledge all interact to create particular distributions of disease at particular places and specific moments in time. The twenty-first-century ecology of sickness in the developed world is marked, for example, by the dominance of chronic and degenerative illness—ailments of the cardiovascular system, of the kidneys, and cancer. What we eat and the work we do or do not do—our physical *and* cultural environment—all help determine our health and longevity.

Disease is historically as well as ecologically specific. Or perhaps I should say that every disease has a unique past. Once discerned and named, every disease claims its own history. At one level, biology creates that identity. Symptoms and epidemiology as well as generation-specific cultural values and scientific understanding shape our responses to illness. Some writers may have romanticized tuberculosis—think of Greta Garbo as Camille—but, as the distinguished medical historian Owsei Temkin noted dryly, no one had ever thought to romanticize dysentery. Tuberculosis was pervasive in nineteenth-century Europe and North America and killed far more women and men than cholera did—but never mobilized the same widespread and policy-shifting anxiety. It was a familiar aspect of life—to be endured if not precisely accepted. Unlike tuberculosis, cholera killed quickly and dramatically and was never assimilated as a condition of life in Europe and North

America. Its episodic visits were always anticipated with fear and tense debates over public health. Sporadic cases of influenza are normally invisible, indistinguishable among a variety of respiratory infections; waves of epidemic flu are all too visible. Syphilis and other sexually transmitted diseases, to cite another example, have had a peculiar and morally inflected attitudinal history. Some diseases, such as smallpox or malaria, have a long history, others, like AIDS, a rather short one. Some, like diabetes and cardiovascular disease, have flourished in modern circumstances; others reflect the realities of an earlier and economically less developed world.

These arguments constitute the logic motivating and underlying the Johns Hopkins Biographies of Disease. *Biography* implies an identity, a chronology, and a narrative—a movement in and through time. Once inscribed by name in our collective understanding of medicine and the body, each disease entity becomes a part of that collective understanding—and thus inevitably shapes the way in which individual men and women think about their own felt symptoms and prospects for future health. Each historically visible entity—each disease—has a distinct history, even if that history is not always defined in terms familiar to a twenty-first-century physician nor tracks neatly onto a modern disease category. Dropsy and Bright's disease are no longer terms of everyday clinical practice, but they are an unavoidable part of the history of chronic kidney disease. Nor do we speak of essential, continued, bilious, and remittent fevers as meaningful designations. Fever is now a symptom, the bodies' response to triggering circumstance; it is no longer a "disease" as it had been through millennia of human history. "Flux," a centuries-old term for diarrhea, is similarly no longer an entity but a symptom associated with a variety of specific and nonspecific causes. The very notion of specific disease entities (fixed and based on a discrete defining mechanism) that we have come to assume is in itself a historical artifact. But it is a very powerful one. We expect a diagnosis when we feel pain or suffer incapacity—expect the world of medicine to at once categorize, explain, predict, and manage.

And for most of history there have been only so many conceptual tools available to make some sense of the otherwise intolerably random imposition of sickness. Constitution has always been one of those explanatory tools: the cards we are dealt by our parents, the tendencies built into us at birth. They are cards that were understood to be differentially played as each woman and man lived through their unique game of life. Tuberculosis and cancer, for example, had often been seen in the past as having a constitutional—that is, inherited—component, influencing though not determining disease incidence. One inherited tendencies, that is, not full-blown ailments; circumstance and lifestyle determined particular outcomes. Before the twentieth century, such notions remained largely impressionistic, just as heredity itself remained an important yet indecipherable black box. And disease itself was only beginning to be seen as a specific *thing*. Research by nineteenth-century pathologists, chemists, and bacteriologists had, over time, helped create this novel understanding of disease as specific, mechanism-based entities. The germ theory and its seeming unraveling of the mystery of infectious disease only underlined the seeming inevitability of this way of understanding disease. If a defining pathological mechanism had not been found to explain a disease, it was only a matter of time before it would be. Even the idiosyncrasy of individual immunity, of resistance and susceptibility, was becoming the subject of systematic investigation and theorizing at the end of the nineteenth century. And heredity, too, figured prominently at the beginning of the twentieth century, as new discoveries promised to shed light on the diversity and ailments of mankind.

By mid-twentieth century, the search for underlying mechanisms of disease had moved to the cellular, chromosomal, even molecular level, as a number of—often rare—diseases became the subject of research that linked biochemical and physical insights with clinical pictures. Sickle-cell disease, for example, was famously christened a molecular disease by Linus Pauling and his coworkers in 1949 and later linked to a particular evolutionarily meaningful mutation. It was in this mid-twentieth-century era of dramatic biomedical innovation that phenylketonuria, or PKU, an uncommon but clinically striking autosomal recessive disease, came to play a highly visible role in shaping public expectations and public health practice in the form of infant screening.

As Diane Paul and Jeffrey Brosco show in the following pages, PKU affirmed the model of a genetic disease as based on a particular misstep in a metabolic pathway; even more significantly in terms of its social impact, the story of PKU promised the power to intervene in an infant's otherwise inevitable downward path toward "mental retardation" and often severe physical and behavioral abnormalities. It was an empowering narrative of discovery and control that appealed widely to anxious parents and interested scientists. If the condition were diagnosed early enough in life, a rigidly controlled diet could avert a tragic and otherwise unavoidable outcome. PKU played a much larger role in public discourse than its comparatively infrequent incidence might imply. But as Paul and Brosco make clear, the story of PKU is cautionary as well as triumphal. Effective management implied a lifetime of careful dietary control. And the widespread adoption of screening for a variety of metabolic diseases in infants has brought therapeutic and diagnostic as well as ethical dilemmas.

Charles E. Rosenberg

PREFACE

Phenylketonuria, or PKU as it is more familiarly known, is a rare disorder, affecting only about one in fifteen thousand individuals. In the United States, for example, about 275 infants are born with the disease each year.[1] Thus, in a lifetime of practice, most physicians will not encounter a single case. Yet, probably every physician in the industrialized world has learned about PKU in medical school, many parents vividly remember the heel-stick test for their newborn, and scientists interested in genetics and metabolism say they hope to "find another PKU." Why has such a rare condition garnered so much attention?

PKU is famous in part because it is widely seen as a victory for scientific medicine. If the condition is detected in the newborn period and a specialized diet is instituted, the profound cognitive impairment usually caused by PKU is averted. For the diet to be effective, however, the otherwise normal-appearing infant with PKU must be identified, among thousands of other nonaffected babies, in the first weeks of life. In the early 1960s, parents of children with mental retardation began to advocate for state laws to test all newborns in the United States. By 1965, thirty-two states had enacted screening laws, all but five making the test compulsory. By the mid-1970s, newborn screening (NBS) for PKU had become routine in nearly every industrialized nation and had even been extended to many poorer countries.

Like many twentieth-century success stories of scientific medicine, the treatment of PKU depended on new methods in both clinical and laboratory medicine, new public health techniques, and new concepts of disease causation and control. As with smallpox, typhoid fever, and pellagra, the control of PKU involved the creation and introduction of a dramatic new method for grappling with a dread disease: in the case of PKU, the mass screening of all newborns. But unlike PKU, other "victories" for modern medicine were over relatively common conditions, either regionally or nationally. Furthermore, appropriate treatment of PKU meant testing millions of unaffected babies, under state sponsorship—an anomaly in health care systems like that of the United States, where government typically has a limited role. Thus our central task in this book is to explain the paradox of PKU: how did a disease of marginal public health significance become the object of an unprecedented system for the routine testing of newborns, and how did it acquire paradigmatic status in the domains of public health and genetics?

Our answer is that PKU is an elusive protagonist, a potent cultural symbol that could be deployed to confront emerging issues in science and medicine. Starting in the 1960s, scientists, politicians, and the general public were excited by the seemingly miraculous outcomes of diet-treated babies with PKU: they viewed this success as an example of the enormous potential of science to transform the lives of people with mental retardation, a cause célèbre in the mid-twentieth century. It seemed that many more examples were sure to follow, with the ability to intervene clinically in PKU portending effective treatments for other cognitive disabilities. Even today, clinicians perform extensive diagnostic testing on people with cognitive disabilities with the hope of finding a simple medical cure, despite the relatively small impact of PKU and similar conditions on the prevalence of mental retardation in the late twentieth century.[2]

In the 1970s and 1980s, PKU took on new roles as a resource for participants both in nature-nurture debates and in controversies over genetic testing. The ubiquity of the case of PKU in both subject domains is striking. So was (and is) its use to illustrate quite disparate points. PKU is a genetic condition that is highly treatable if infants are diagnosed at birth and placed on a diet low in phenylalanine, an essential amino acid found in all dietary proteins. That an environmental intervention could dramatically alter the course of a genetic disorder made it an attractive example both for critics of genetic determinism, who employed it in both the genetics of IQ and the sociobiology controversies, and for enthusiasts for genetic screening, who traded on the "large store of goodwill and ethical credit" accumulated by NBS to legitimate genetic screening programs more generally.[3] Foes of genetic determinism and promoters of genetic tests have not generally been political bedfellows and, indeed, are often at odds; yet both find support for their positions in the history of PKU.

In the 1990s and 2000s, PKU served as the paradigmatic example for advocates eager to expand newborn screening. Twenty-first-century NBS programs can potentially identify dozens of conditions, and proponents of expansion continue to use PKU to underscore the value of screening infants for disease. However, advocates inherit not just the goodwill generated by the success of treatment for PKU but also the challenges: NBS programs were established a half-century ago in specific historical conditions that do not necessarily obtain today. For example, NBS programs are almost everywhere, de facto or de jure, mandatory, with no provisions for informed consent. Many professionals and parents' organizations would now like to test for conditions that are much less treatable than PKU or for which testing provides no direct medical benefit at all to the child. Moreover, the dried blood spots collected from millions of newborns, representing an unselected population, have become a valuable resource for researchers. But these potential clinical and research practices, which would seem to require specific informed consent from parents, are not easy to harmonize with a legal framework and hospital routines that date from the 1960s.

The regularity with which PKU was and is invoked in the service of divergent agendas prompts an obvious question: how well does this ubiquitous PKU success story accord with reality? Accounts of PKU diagnosis and treatment in the nonspecialist literature rarely mention any diagnostic complications, treatment challenges, or imperfect outcomes. Assumptions about the ease of adhering to the low-phenylalanine diet tend to be particularly naive, as exemplified by a 2012 *Nature* article asserting that treatment for newborns with PKU and other metabolic and immunological disorders "is often as simple as diet change."[4] But, as we will see, there is nothing simple about adhering to the PKU diet.

A common analogy has been between the low-phenylalanine diet and insulin, as in a 1964 *New York Times* report on the new law mandating PKU testing: "The disease can be detected by a simple blood test costing about 50 cents and can be corrected by a special diet in much the same way as diabetics are enabled to lead normal lives by the use of insulin."[5] It turns out that PKU is indeed quite like type 1 diabetes, though not as the *Times* writer intended. As anyone with either a personal or a professional interest in diabetes—or PKU—knows, therapy for these conditions involves a lifelong struggle with personal behaviors and medical management. In this way, one strand of the history of PKU follows the contours of the discovery of insulin described by historian Chris Feudtner in *Bittersweet*: the famous breakthrough of scientific medicine does indeed save lives and reduce morbidity but only through the arduous and uncertain path of living with a chronic condition.[6]

In the end, PKU will always be remembered as the first condition for which newborns were screened. But its continuing value in policy debates is explained by the paucity of more powerful examples of the success of modern scientific medicine in preventing mental retardation, in altering the course of genetic disease, or in dramatically changing outcomes through NBS. Despite great promise over the past fifty years, there have been precious few victories in these realms comparable to the success of testing and treatment for PKU. That is why it continues, even today, to serve as an exemplar for diverse constituencies.

❖ ❖ ❖

Despite PKU's singular role in the histories of human genetics and of public health and its cultural importance, this is the first book-length treatment of the condition by historians. This book traces the history of PKU from its discovery, in the 1930s, as a metabolic disorder through the development of newborn screening programs and their expansion. It asks how scientific, technological, economic, cultural, and political forces shaped responses to PKU and, conversely, how the nature of these responses might enrich or sometimes challenge conventional interpretations of social and scientific trends. But this short book does not aim to be comprehensive. In particular, we have tried to include sufficient technical detail to satisfy scientists, clinicians, and others interested in the biochemistry and mode of inheritance of PKU without overwhelming other readers. We hope the latter will include not only historians of science and medicine and bioethicists but anyone concerned with issues in genetics and, more generally, medical screening, as well as individuals who live with PKU and their families.[7]

The story we tell has both US and international components. Several chapters focus on the American experience, reflecting not only the authors' expertise but the fact that the Guthrie heel-stick test that made universal newborn screening possible was invented in the United States. The United States was also the sole country where screening was mandated by law—and passionately contested. How proponents of compulsory screening were able to prevail in a country where organized medicine has historically been hostile to government medical programs is a central theme of this book.

We have tried, however, to do justice to the international dimensions of the history. After all, PKU was first described as a disease entity by a Norwegian, and much of the early research on the biochemistry and genetics of PKU and on the development of a dietary therapy was British. Moreover, the first programs to screen newborns for PKU were established in the United Kingdom, not, as is often assumed, the United States. Elsewhere, newborn screening was, initially, largely an American export, promoted through professional networks and US government programs. But eventually, the trajectory of screening in the United States and in much of the rest of the world would come to diverge. In most US states, the system originally designed to detect PKU would be expanded to include many other conditions detectable in newborns, whereas in most European and other countries, that system would expand only slowly and modestly. We aim to identify the technological, commercial, political, and institutional factors that explain why, in contrast to the United States, most countries have adopted a very cautious approach to expansion of NBS.

Apart from the broad themes outlined above, we think it worth remarking on several other aspects of the PKU story. One that surprised us was the degree of enthusiasm for newborn screening that existed before Robert Guthrie's invention of an assay that could reliably detect PKU in newborns. Today, the invention of a PKU screening test is almost invariably associated with Guthrie, and most accounts of the history of screening (including Diane Paul's own earlier work) begin in the 1960s. Yet urine screening for PKU began in several countries, including the United States, the United Kingdom, and Spain, in the 1950s, and the prospect of detecting and treating the condition in infancy generated enormous excitement and even efforts to make screening mandatory by law. The extent to which Guthrie was able to exploit an already existing momentum for screening is an underappreciated aspect of the context for his success.

Another surprise is the extent to which the histories of PKU and eugenics intersect. The standard view is that newborn screening "marked a sharp departure from the earlier association of human genetics with the eugenics movement."[8] For some individuals and organizations it certainly did. After all, PKU first came to public prominence in 1946, when it was employed by the British polymath Lionel Penrose to illustrate what was wrong with eugenics—which was also no part of the agenda of Robert Guthrie and the National Association for Retarded Citizens (NARC; originally named the National Association for Retarded Children). It thus seems eminently reasonable to counterpoise treatment for PKU, which allows the responsible genes to spread, with eugenics, which aims to reduce the incidence of such genes through selective breeding.

The story is not so simple, however. All the early prominent PKU researchers, including critics of eugenics such as Penrose, hoped to develop methods to detect carriers of the condition so that heterozygotes could avoid marrying each other. Even following the advent of widespread treatment for PKU in the 1960s, concerns about carriers lingered in some quarters. Furthermore, as late as the 1980s, many young women with PKU were warned not to get pregnant and to abort if they did, so as to avoid giving birth to a baby harmed by exposure to phenylalanine circulating in maternal blood. Thus the history of PKU sensitizes us to the ways in which the meaning of terms such as "eugenics" has shifted without our noticing, such that many of yesteryear's critics would appear by today's standards to be advocates of eugenics.

Another notable feature of the history is the degree to which universal NBS was adopted in the United States and elsewhere *before* there was widespread consensus on a series of fundamental scientific and policy issues. Well into the late 1960s—and for some issues, decades later—researchers and clinicians disagreed about whether the Guthrie test was reliable, sensitive, or specific enough for population-based screening; who should be treated and for how long; whether dietary treatment worked and, if so, whether the risks of treatment outweighed the benefits; and even whether an elevated phenylalanine blood level was itself the cause of the cognitive impairment. In the 1960s, organized medical groups such as the American Academy of Pediatrics and the American Medical Association opposed government-mandated screening for PKU, as did experts from fields as diverse as public health and human metabolism. In the end, the skeptics had little impact: they were simply no match for the well-organized and politically savvy parent volunteers and professionals who lobbied state governments under the leadership of the NARC.

❖ ❖ ❖

The two authors of this history came to the topic by different paths. Diane Paul's curiosity about phenylketonuria was piqued in the early 1990s, when she was engaged in research on the history of the nature-nurture debate and on policy issues around genetic testing. This book represents the fruit of an almost twenty-year effort by Diane both to understand how PKU came to acquire its symbolic power and to identify the facets of the real-world experience that this process has rendered largely invisible. Diane did the archival research, conducted the oral history and patient interviews, and wrote the first draft of nearly every chapter. Jeff Brosco came to the subject of PKU first as a developmental pediatrician and later as a historian interested in PKU as a paradigmatic example of modern medicine's enthusiasm for science as the best path for confronting mental retardation.[9] He drafted chapter 2 and contributed substantially to the organization and content of the other chapters, adding his perspective as someone involved in developing health policy for newborn screening. Together, we developed the book's key themes. We believe that the collaboration of scholars with different backgrounds, training, and orientations toward screening has produced a richer and more balanced account of the PKU story than would otherwise have been possible.

A NOTE ON LANGUAGE

Given the close relationship between PKU and cognitive impairment and the politically charged nature of language, we would like to briefly address how we chose to describe people with cognitive impairment in this book. In the United States in 2013, we would describe a person with untreated classical PKU as having an "intellectual disability." That is, the moderate to severe forms of PKU would lead to differences in the brain that cause a significant degree of cognitive impairment, as measured by a trained professional using a standardized test of intelligence (e.g., a score more than two standard deviations below the mean for a population). If the degree of impairment caused an individual significant functional difficulty in his or her specific environment (e.g., school), then an "intellectual disability" would be diagnosed.[10] This new US definition is consistent with the World Health Organization's *International Classification of Functioning, Disability, and Health*, which emphasizes that an individual's impairment is mitigated or exacerbated by the environment; "disability" is thus a description of the interaction between individual and environment.[11]

"Intellectual disability" is a relatively new term that means exactly the same as the most modern meaning of the words "mental retardation." After exhaustive research into terminology,[12] as well as careful deliberation regarding the legal and social impact of changing the terminology, in 2007 the American Association of Mental Retardation changed its name—and the title of its authoritative guide—to the American Association on Intellectual and Developmental Disabilities. Although "mental retardation" is still widely used in the United States, many advocates find the term pejorative and are working to change the language to "intellectual disability" in law and medicine, as well as in more general usage.

This is only the latest, and surely not final, change.[13] Over the past two centuries, different words have been used to describe a person with a moderate to severe cognitive impairment, and indeed, those words have different meanings specific to time and place: from "idiot" and "imbecile" of the nineteenth century to the addition of "feebleminded" and then "moron" by the early twentieth century. In the mid-twentieth century, "mental retardation" was considered progressive and nonstigmatizing. Because these words describe a political and social response as much as an individual impairment, in this book we have endeavored to use the term appropriate for the specific era being discussed.

ABBREVIATIONS

AAP	American Academy of Pediatrics
ACMG	American College of Medical Genetics
AMA	American Medical Association
BIA	bacterial inhibition assay
CH	congenital hypothyroidism
FDA	US Food and Drug Administration
GMP	glycomacropeptide
HEW	Department of Health, Education, and Welfare
MCADD	medium-chain acyl-CoA dehydrogenase deficiency
MPKU	maternal PKU
MPKUCS	Maternal PKU Collaborative Study
MPKUS	maternal PKU syndrome
MR	mental retardation
MS-MS	tandem mass spectrometry
MSUD	maple syrup urine disease
NARC	National Association for Retarded Children (1953–73); National Association for Retarded Citizens (1973–81)
NAS	National Academy of Sciences
NBS	newborn screening
NICHD	National Institute of Child Health and Human Development
NIH	National Institutes of Health
PAH	phenylalanine hydroxylase
phe	phenylalanine
PKU	phenylketonuria
PKUCS	United States Collaborative Study of Children Treated for Phenylketonuria

PPMR	President's Panel on Mental Retardation
rDNA	recombinant DNA
SACHDNC	Secretary's Advisory Committee on Heritable Disorders in Newborns and Children
SCD	sickle-cell disease
USCB	United States Children's Bureau
WHO	World Health Organization

THE PKU PARADOX

INTRODUCTION

Pearl Buck, PKU, and Mental Retardation

In May 1950, Pearl S. Buck became the first prominent American to acknowledge having a child with mental retardation. Her essay "The Child Who Never Grew," which first appeared in the *Ladies Home Journal* and was soon expanded into a small book, told the story of her cognitively impaired daughter, Carol. At the time, Carol was 30 years old and living at the Vineland Training School in New Jersey, where she had been resident for more than two decades.[1] Carol had been a beautiful blond and blue-eyed baby; strong and healthy, at first she seemed to develop normally. But she was slow to talk, and her mother began to grow uneasy. Although friends and others whom she informally consulted were reassuring, there were mounting signs of trouble. Carol exhibited not only delayed speech but also a very short attention span, and her movements often seemed purposeless. She also developed eczema, and her urine had a notably strong, musty odor. By the time Carol was 4 years old, her mother was forced to acknowledge that something was seriously amiss. A long and frustrating diagnostic odyssey ensued, first in China, where the family was then living, and later during a visit to the United States, where "the end of each conference was to send us on to someone else, perhaps a thousand miles away."[2] Eventually, a physician at the Mayo Clinic informed Buck that, for some unknown reason, Carol's mental development had apparently ceased.

Mother and daughter returned to China, but the country was soon wracked by civil war and revolution, with the Bucks at one point only narrowly avoiding death when they were hidden by a peasant neighbor.[3] Pearl Buck, anxious about Carol's fate should her mother die prematurely and finding it increasingly difficult to manage her at home, decided to place her daughter in a residential facility. But even if the political situation had been less chaotic in the 1920s, there were no such institutions in China, where the disabled were cared for at home. Thus Pearl Buck and her husband, John Lossing Buck, decided to find a school for their daughter in the United States, and in 1929, Buck traveled there with the now 9-year-old Carol and her younger adopted sister, Janice.[4]

The choices were mostly grim. As the family was not resident in the United States, Carol was ineligible for most state schools, which tended, in any case, to be overcrowded, with the children living lives of strict routine. Some expensive private institutions, although physically handsome and well equipped, seemed no better from the perspective of resident care. A bright spot in this generally depressing landscape was Vineland, a private institution that was organized into homelike cottages and had a compassionate director and staff. But Vineland cost $1,000 a year, a fee that Buck was able to afford only with a loan from a patron.[5] It was to help pay for Carol's care that she wrote *The Good Earth*, her 1931 Pulitzer Prize–winning novel of life in China; its proceeds ultimately enabled her to establish a substantial endowment for Carol at Vineland. Buck would become an active member of Vineland's board of directors and, as she achieved financial success, an important donor, assigning all profits from *The Child Who Never Grew* to the institution and endowing an attractive two-story cottage, with a porch and wading pool, for Carol and 15 other children.[6] Carol's suite included a large bedroom furnished with French Provincial furniture, a small bath, and a playroom with windows on three sides.[7] Carol is buried at Vineland, where

Passport photo of the Buck family; *left to right*, Janice, Lossing, Carol, and Pearl.
Courtesy of Pearl S. Buck International, www.pearlsbuck.org.

she lived until her death in 1992 and where she was visited by both her parents.

In some ways, Carol's story is similar to that of many thousands of children with mental retardation in the first half of the twentieth century. Although Pearl Buck was certainly atypical with respect to the financial resources she was ultimately able to devote to her daughter's care, she faced the same social stigma of mental retardation, the same limited scientific understanding of the causes and treatment, and the same deeply personal struggles over how best to meet the needs of a child in the midst of difficult family circumstances. Almost three decades after entering Vine-

Carol Cottage, Vineland, NJ.
Courtesy of Donna Rhodes, Curator, Pearl S. Buck National Historic Landmark Home.

land, and long after she might have benefited from treatment, Carol was diagnosed with phenylketonuria (PKU).[8]

THE METABOLIC AND GENETIC BASIS FOR PKU

Carol had many of the characteristic manifestations of PKU: cognitive impairment, microcephaly (small head size), eczema, a clumsy gait, hyperactivity, seizures, behavioral disruptions, and a strong "musty" odor to the urine. Because melanin synthesis is disrupted in PKU, she also had the typical blue eyes, blond hair, and fair skin. These traits result from the failure of the body to process phenylalanine, an essential amino acid that is necessary for protein synthesis and normal growth and development. Humans do not produce phenylalanine endogenously but must obtain it from the foods they eat (hence its designation as an "essential" amino acid). Only a portion of the phenylalanine ingested in the diet is required for the synthesis of new proteins and other functions; ordinarily, the remainder is converted to another amino acid, tyrosine. As a result of a mutation inherited from both parents, people with PKU have little or no activity in a liver enzyme called phenylalanine hydroxylase (PAH), which catalyzes the conversion of phenylalanine to tyrosine. When the normal metabolic pathway is blocked, phenylalanine and its metabolites accumulate in the blood and body tissues, and in some (still unknown) way impair normal development of the brain of the newborn. The damage occurs after birth, since in utero the fetus is protected by the mother's metabolism, which clears the excess phenylalanine.

Although PKU is often described as a classical, "simple Mendelian disorder," nothing about the disease is simple, including its genetics. More than five hundred mutations have been identified at the PAH gene locus, and this genetic heterogeneity is associated with considerable clinical heterogeneity.[9] Most people diagnosed with PKU are "compound heterozygotes"; that is, they have inherited two different mutations in the PAH gene from their parents. Thus, PKU can be mild, moderate, or severe, statuses with different implications for treatment.[10] In severe cases, where there is no or almost no enzyme activity and thus a high concentration of phenylalanine in the blood (what came to be called "classical PKU"), untreated disease is associated with profound cognitive impairment. The child may never learn to walk or talk or to control his or her bowels or bladder.

THE STIGMA OF MENTAL RETARDATION

Before "The Child Who Never Grew" was published in 1950, Buck did not speak publicly, or for the most part even privately, of her adult daughter's existence. Whether to protect Carol's privacy or to spare herself pain and embarrassment, Buck admonished her few confidantes to keep this knowledge to themselves. In interviews, Buck would at most say that she had two daughters, one of whom was away at school. Why she ultimately decided to tell her story to the world is not completely clear, but at least one factor seems to have been the ferment created by conscientious objectors' reports of appalling conditions in state custodial institutions and by a related 1948 exposé published by the newly formed National Mental Health Foundation, on whose board Buck served.[11] Thus, in explaining why the time had finally come for her to speak out, Buck wrote that she wanted her "child's life to be of use in her generation" and that "there is afoot in our country a great new movement to help all children like her."[12] Buck may also have been referring to the emerging local parents' organizations, which by the late 1940s included thousands of family members across the United States who were advocating for their children with mental retardation to have access to local schools and recreational activities.[13] In any case, *The Child Who Never Grew*, which sold for $1, clearly touched a chord with many parents. Translated into 13 languages, it was summarized in *Time* magazine, excerpted in the *Reader's Digest*, and generated "mailbags of letters from readers."[14] It also prompted other parents—some celebrities like Buck, others not—to write of their own efforts to come to terms with the fact of having a profoundly disabled child.

The difficulty of those struggles was greatly exacerbated by the stigma that attached to mental retardation. In the first decades of the twentieth century, what was then often called "feeblemindedness" was associated with stigmatized populations, especially urban immigrants and rural clans; furthermore, it was thought to be the root cause of many social problems, including crime, poverty, substance abuse, and sexual promiscuity. This last association was particularly critical, because feeblemindedness was noted to run in families, and a genetic etiology was widely assumed. Indeed, many American scholars believed that mental defect was inherited as a simple Mendelian recessive, and as such, many of society's greatest problems—linked to feeblemindedness—could be eliminated by applying the lessons of animal breeding to humans. (The study that seemed to confirm the inherited nature of feeblemindedness, H. H. Goddard's *The Kallikak Family: A Study in the Heredity of Feeble-Mindedness*, published in 1912, chronicled the family history of a Vineland resident.) Americans across the political spectrum found common ground in eugenics, the effort to promote childbirth among the healthy, intelligent, and responsible and discourage childbirth among those likely to produce children with mental, moral, and physical defects. The occasional appearance of feeblemindedness in prosperous families of good character was understood as a genetic phenomenon, probably related to the transgression of some distant ancestor. For this reason, eugenicists generally counseled even the well-off to investigate the family lines of potential spouses and avoid marriage into lines with this taint. No wonder that through the 1930s and 1940s, mental defect in a relative was often treated as a shameful secret.

Buck tried to undermine that stigma by emphasizing that mental defect was *not* usually inherited and therefore not the family's fault. In *The Child Who Never Grew*, she repeatedly asserted that most of the mentally retarded are so affected due to noninherited causes and that "the old stigma of 'something in the family' is all too often unjust.'"[15] Her insistence that there was absolutely no sign of trouble in her own family history is thus understandable. It is also ironic, as she finally learned in the late 1950s that Carol had PKU. Phenylketonuria was understood to be an autosomal recessive disorder (meaning that the mutation occurs on a gene in the non-sex chromosomes, or autosomes, and that a copy of the abnormal gene must originate from each parent). Buck was thus aware that both she and her former husband had to be carriers, and indeed, she wrote to Lossing to tell him the news, for the sake of the two children from his second marriage.[16] Clearly very proud of her lineage, Buck had noted in *The Child Who Never Grew* that she "was fortunate in her own ancestry on both sides" and that there was absolutely nothing in her family history to have made her concerned about the possibility of mental retardation. Although she desperately wished to know the cause of her daughter's condition, the discovery that it was inherited must have come as a shock; her adopted daughter noted that she "had trouble accepting that her family's genes may have contributed to this disorder."[17]

In a book that aimed to give parents hope, there was another reason for Buck to stress that most mental retardation was not inherited. In 1950, to identify a trait as inherited not only was stigmatizing but implied that nothing could be done to alter its course. Given the assumption that inherited meant intractable, the only method for preventing the appearance of traits that ran in families was to prevent those likely to be affected from being born in the first place; that is, through eugenic segregation or sterilization. The fatalism that attached to a genetic etiology is reflected in Buck's admonition, "Let us remember that more than half of the mentally deficient in this country are so from noninherited causes, and these causes can be prevented, did we know what they are."[18]

THE ROAD TO NEWBORN SCREENING

In the 1950s, a treatment was developed for phenylketonuria, thus demonstrating that inherited mental retardation and, indeed, the damaging effects of other inherited conditions might, after all, be prevented through medical intervention. Experimental low-phenylalanine diets were shown to ameliorate at least some effects of the disease—but also to be most effective when initiated in early infancy, before symptoms developed. Those experiments converged with the development of a urine test that could be used to detect the disease in infants and a new focus on the scientific prevention of mental retardation. Some physicians, hospitals, and public health programs in both the United States and the United Kingdom began to screen asymptomatic newborns. But screening really took off in the 1960s with the development of a much more reliable and inexpensive blood test that could be administered just a few days after birth, before brain damage occurred.

In the United States, parent groups frustrated by what they considered the slow uptake of screening lobbied state legislatures to make the test mandatory. Their efforts were opposed by state medical societies worried about malpractice suits and state intrusion in the doctor-patient relationship, but also by experts concerned that not enough was known about the reliability of the test, the natural history of the disease, or the safety or efficacy of treatment to justify compulsion. The medical societies and researchers proved no match politically for the highly motivated and organized parents and their allies, and by the late 1960s, most US states had mandated PKU screening in newborns. In the United Kingdom, Canada, Australia, New Zealand, and most countries in Western Europe, screening was rapidly routinized, even though not required by law. Today, newborns in much of the world are tested for this disease.

The ability to intervene in PKU is now considered to be a seminal moment in medicine and to have "constituted a 'paradigm shift' in medical thinking about genetic disease in general."[19] Biochemical geneticist Charles Scriver notes that, prior to the advent of early diagnosis and treatment of PKU, genetic diseases elicited a "yawn" from the medical profession. "Medicine is supposed to heal and treat. And all those genetic things? The die is cast, so what can you do about it?"[20] But the PKU experience showed that such diseases were not, in fact, unalterable. As one researcher reflected in 1982:

> In establishing neonatal screening for PKU, the legislators and officials in numerous states have implemented one of the first large-scale efforts in preventing the consequences of genetic diseases. These programs and their early success have been widely cited in textbooks of biology and genetics and in lectures to the general public. There is something appealing to the human spirit in the ability to thwart the mental retardation of PKU. It is one victory in the struggle against genetic factors, which are seen as being unalterable. PKU programs have become a showcase of the benefits to be derived from large-scale screening for genetic disorders.[21]

The history of that shift in thinking begins in 1930s Norway, with an encounter between another distraught but tenacious mother and the biochemist-physician who reluctantly agreed to examine her children.

CHAPTER ONE

The Discovery of PKU as a Metabolic Disorder

A NEW "INBORN ERROR OF METABOLISM"

Borgny Egeland gave birth to her first child, a daughter named Liv, in Norway in 1927. Like Carol Buck, Liv Egeland seemed fine in early infancy but was slow to begin to talk. Consulting the family doctor, this mother, too, was assured there was no cause for concern. A brother, Dag, born when Liv was 3 years old, also seemed healthy at birth but would soon develop symptoms more severe than his sister's. Liv learned to speak a few words and exhibited a spastic gait and seemingly random movements; in contrast, Dag gradually lost interest in his surroundings and was never able to talk, walk, chew solid food, or even sit up by himself. Noticing a strong musty odor to the body and urine of both children, which could not be eliminated no matter how often they were bathed, the parents wondered whether it might be related to their developmental problems. But neither their family physician nor any of a host of other doctors, herbalists, and other healers could identify either the strange odor or the cause of their children's ills.

Colleagues of Borgny Egeland's husband, Harry, a dentist, suggested that the parents contact Ivar Asbjørn Følling (1888–1973), a physician and biochemist with a strong interest in metabolic disease. As it turned out, Borgny Egeland's sister occasionally saw

Følling socially and so was asked, when an opportunity arose, to tell him about the children's retardation and inquire whether he thought it might be linked to their peculiar musty odor. Følling later explained that, although he had no real expectation of being able to help, he did not want to disappoint the mother and so offered to examine Liv and Dag, if she would bring the children and a urine sample to his laboratory.[1]

That fortuitous connection would have far-reaching consequences. Følling was one of the few Norwegians well trained in both chemistry and medicine, and as such, he was well positioned to appreciate that diseases could result from failures of metabolism, could be identified biochemically, and could be inherited as Mendelian recessives. In 1927–28, Følling had been awarded a Rockefeller Foundation fellowship that funded study at Harvard University, where he worked with biochemist-physiologist Lawrence J. Henderson. After further studies at several prominent US institutions, including work at the Rockefeller Institute with Donald Van Slyke, one of the founders of clinical chemistry, Følling returned to Norway to assume a faculty position at the University of Oslo.[2]

As an academically oriented biochemist, Følling was familiar with Archibald Garrod's concept of "inborn errors of metabolism" (what we would now refer to as inherited metabolic disorders), an expression Garrod coined to describe a group of conditions that "apparently result from the failure of some step or other in the series of chemical changes that constitute metabolism."[3] In his landmark 1902 paper "The Incidence of Alkaptonuria: A Study in Chemical Individuality," a copy of which was later found in Følling's attic, Garrod claimed that the condition known colloquially as "black urine disease" was due, not to intestinal microorganisms, but to a more or less harmless chemical aberration whose pattern of transmission could be accounted for by Mendel's laws of heredity.[4] More generally, Garrod argued that in their chemistry as much as in their physical structure, no two individuals of a species are identical. Most differences, whether physical or chemical, are slight, but some are very marked—as in the case of

Ivar Asbjørn Følling (*front*) in a mountain cabin, Colorado, 1930.
From S. A. Centerwall and W. R Centerwall, "The Discovery of Phenylketonuria,"
Pediatrics 105 (2000), courtesy of John Wiley & Sons, Ltd.

supernumerary digits or the metabolic condition alkaptonuria.[5] In the Croonian Lectures delivered to the Royal Society of Physicians in 1908 and published the following year as *Inborn Errors of Metabolism*, Garrod identified three other such "metabolic sports" linked to enzyme deficiencies—albinism, cystinuria, and pentosuria—and demonstrated that they were inherited as Mendelian autosomal recessives.

When the Egelands brought their children to Følling's lab, he applied his skills as a chemist as well as a physician. As a routine part of the examination, he added a few drops of ferric chloride to the children's urine, at that time a standard test for the detection of ketones in diabetics.[6] To his surprise, the urine turned a deep green color. Such a reaction had never before been described, and after confirming that it was not a fluke, he set out to identify the causative substance. The process of separating the unknown compound from the thousands of others in the urine and then purifying it was laborious, requiring 22 liters of the children's urine, but Følling ultimately determined that the substance was phenylpyruvic acid (which was not, however, the source of the offensive smell).[7]

Once the acid was isolated, the next step was to ascertain whether the Egeland children were unique in excreting it. Følling analyzed urine samples from 430 mostly institutionalized patients and found that 8 tested positive. In a 1934 paper, Følling reported that of the 10 patients who excreted the acid, including the Egeland children, 9 were definitely feebleminded, while the tenth was too young to be diagnosed. Given that the most prominent symptom of the disease was cognitive impairment and that it was identified by the presence of phenylpyruvic acid in the urine, he called this newly discovered disease "imbecillitas phenylpyruvica."[8]

Structurally, phenylalanine and phenylpyruvic acid are very similar (differing only by the substitution of an amino group by an oxygen atom), indicating that the two compounds were almost certainly metabolically related.[9] Følling also knew that rabbits fed excessive amounts of phenylalanine excreted the acid, as had one of his patients when fed a phenylalanine "load." He thus hypothesized that the acid was produced when the normal metabolism of phenylalanine was blocked, resulting in deflection of the metabolic process into an abnormal pathway. At the time, there was no satisfactory way to measure phenylalanine in the blood or urine, so Følling and his colleagues invented a test to verify that affected individuals had markedly elevated levels of phenylalanine. They then proceeded with loading experiments to demonstrate that adding phenylalanine to the diet of animals and affected humans resulted in an increase in the excretion of phenylpyruvic acid.[10] Since phenylpyruvic acid is not a normal breakdown product of phenylalanine and was thought never to be present in the urine of healthy persons, it seemed plausible to Følling that the metabolic anomaly was in some way related to the impairment in phenylpyruvic acid metabolism.[11] (Følling can be seen discussing his discovery in an interview filmed in the 1970s.)[12]

Several facts suggested to Følling that the condition might be inherited. Among the 10 patients who had tested positive for the presence of phenylpyruvic acid were three pairs of siblings. The families of the 10 patients also included three consanguineous marriages, and two of the parents had a total of 12 children in second marriages, all of whom were healthy. This pattern suggested that the disease was inherited as an autosomal recessive, a result confirmed in collaborative research with colleagues.[13]

Følling's original paper on PKU caused little stir at the time, but it did motivate other researchers to look for cases of the disease. Among them were the American physician and biochemist George A. Jervis and the British biochemical geneticist Lionel S. Penrose.[14] In 1935, Penrose, then research medical officer at the Royal Counties' Institution, Colchester, tested the urine of five hundred patients and reported two positive cases.[15] In a second paper, Penrose presented a pedigree analysis that implicated a single recessive gene as the cause. However, Penrose thought that the gene was incompletely recessive, hypothesizing that carriers were disposed to mental breakdown. He also noted that although homozygotes were rare, carriers were not, and that the latter might account for a sizable percentage of the mentally disordered.[16]

At the suggestion of his biochemist collaborator Juda Quastel, Penrose proposed replacing Følling's "imbecillitas phenylpyruvica" and Jervis's "phenylpyruvic oligophrenia" (the expression favored in the United States) with the term "phenylketonuria." He reasoned that a conspicuous feature of the disease was the presence of phenylpyruvic acid, a phenylketone, and that the new term would be consistent with the nomenclature used by Garrod for other inborn errors of metabolism in which the excretion of an abnormal substance was the most striking feature of the condition, as in alkapton and alkaptonuria, or cystine and cystinuria.[17] (In Norway, the condition is still known as "Følling's disease.") In his unpublished memoir, Penrose commented resentfully that "our efforts were eventually rewarded by the invention of an abominable abbreviation, PKU, which arose in America and has now spread all over the world."[18]

Although Penrose did much to publicize the disease and make it a powerful anti-eugenic symbol, the most significant work to clarify both its biochemistry and its mode of inheritance was conducted by Jervis, director of research at Letchworth Village Hospital in New York. In 1939, Jervis analyzed 213 institutionalized patients and confirmed Penrose's supposition that the condition was due to a single autosomal recessive gene. He also confirmed, as Følling had surmised in 1934, that it was associated with an excess of phenylalanine, and he was able to identify the cause as an inability to convert phenylalanine to another amino acid, tyrosine. Eventually, Jervis would demonstrate that this inability results from the absence of a liver enzyme, phenylalanine hydroxylase.[19]

Given that the disease was associated with an excess of phenylalanine, it soon occurred to researchers that the disease might be treatable if a practical way could be found to exclude most phenylalanine from the diet of susceptible individuals. Indeed, the first physician to successfully prepare and administer a nutritional therapy later remarked that "the idea of treating phenylketonuria with a diet low in phenylalanine probably occurs to everybody who studies the biochemistry of the disease."[20]

But what seemed straightforward in theory was complicated by uncertainty about the exact pathogenesis of the disease, which was (and indeed remains) contested, and by practical obstacles to the creation of an effective and affordable diet.[21] Penrose, for example, conducted experiments based on the simple strategy of removing normal protein sources from the diet, but his efforts foundered when a diet consisting of fruit, sugar, olive oil, dripping, jellies, and vitamin pills resulted in severe malnutrition. When he then consulted Sir Frederick Gowland Hopkins, Britain's leading biochemist, about the prospect of constructing a diet that would be nutritionally adequate, he was told that it would cost about £1,000 per patient per week.[22] Penrose was further discouraged by his uncertainty that the cognitive impairment in phenylketonuria was actually caused by the excess of phenylalanine or its abnormal metabolites. By the end of the 1930s, Penrose's experiments with nutritional therapy lapsed.

In the United States, Jervis and Richard J. Block announced plans to develop a low-phenylalanine diet at a 1939 meeting of the American Chemical Society.[23] Block was author, with Diana Bolling, of *The Amino Acid Composition of Proteins and Foods*, the bible for laboratory procedures for amino acid and protein research. He recognized that much of the phenylalanine in a protein hydrolysate (the mixture of amino acids produced from breakdown of a protein) would be removed if the mix were run through a column of activated charcoal, to which the phenylalanine would stick. Although Block's work with Bolling would later prove crucial in efforts to develop a workable diet, in 1939 Block was unable to interest authorities on mental retardation in his work and thus could not get research funding for facilities to make the preparation.[24] It was not until the 1950s that efforts to create a low-phenylalanine diet succeeded, as a result of independent efforts in Britain and the United States (a process described in chapter 3).

PROBLEMS OF INHERITANCE

Accounts of the history of research on PKU invariably describe a trajectory in which an understanding of the biochemistry prompted an interest in dietary interventions that finally bore fruit. That story is true as far as it goes, but it ignores some important detours. One relates to the understanding of the disease's mode of inheritance. Indeed, researchers' main interest in the two decades following the discovery of PKU was not in developing a treatment but rather in detecting heterozygotes, or carriers. Why did it seem important to identify carriers?

The answer in brief is that carrier detection was considered a precondition for any practical and effective program to control the disease. Dietary treatment, while theoretically intriguing, had gone nowhere in practice, and some researchers doubted that it ever would. But there was considerable optimism about the possibility of identifying close relatives of individuals with PKU, who could then be discouraged from reproducing. Reflecting back on the discovery of PKU, Følling wrote, "Early during my investigations in this field I tried to find a way of selecting the heterozygote carriers. This seemed to me to be very important, because it might lead to prevention of those marriages that could produce phenylketonuric patients."[25] George Jervis had a similar aim. After noting that the disease was illustrative of a type of mental deficiency determined by a single autosomal recessive gene, he went on to say, "The practical implications of this conclusion are obvious. Parents of children affected with phenylpyruvic oligophrenia should be discouraged from having other children. Parenthood should be also discouraged in brothers and sisters, uncles and aunts of affected individuals. Consanguineous marriages among members of families of patients should particularly be prevented. Moreover, the patients should be segregated to prevent childbearing, since the great majority reach sexual maturity."[26]

Lionel Penrose was similarly motivated in the 1940s to work on developing a test for identifying heterozygotes (based on the assumption that the urine of carriers would contain a small amount of phenylpyruvic acid).[27] Penrose's interest in carrier detection might seem surprising given his deserved reputation as one of the foremost critics of eugenics in Britain.[28] His arguments against eugenics were both ethical and scientific. Penrose maintained that the best index of a society's health was its willingness to provide adequate care for those unable to care for themselves and, in his role as a leading expert on the genetics of mental deficiency, that the causes of such deficiency were heterogeneous and the impact of eugenic measures on reducing its incidence were modest, at best. On this last point, Penrose is known particularly for his use of PKU to illustrate the futility of efforts to select against deleterious genes. Understanding the nature of his interest in carrier detection in the context of his wider views thus illuminates not just the history of PKU but changing understandings of the meaning of eugenics.

In 1945, Penrose was appointed Galton Professor of Eugenics and director of the Galton Laboratory for National Eugenics at University College London. His inaugural address, delivered in January 1946 and published in the *Lancet*, was titled "Phenylketonuria: A Problem in Eugenics." It was a slashing critique of the scientific and social assumptions associated with Galtonian eugenics, especially in Britain, the United States, and Germany. In his recent book on the history of human genetics, Nathaniel Comfort notes that many biochemical geneticists considered the address "a manifesto of the replacement of eugenics by modern medical genetics."[29] In the immediate aftermath of the Nazi defeat, Penrose argued that eugenics was not only immoral but ineffectual. He used the case of PKU to illustrate why.

The problem referred to in Penrose's lecture title is recessivity, a property of the gene that retards the process of selection against a trait. Since at least the early 1920s, geneticists had understood that when a gene is both recessive and rare, it will be maintained in the population almost entirely through reproduction by apparently normal carriers. Targeting those who are affected (i.e., who exhibit the trait) will only slightly reduce their number, which will be constantly replenished from the large heterozygotic reserve. In the case of PKU, there is an added consideration: affected individuals are almost always infertile. Thus, selection aimed at them would have virtually no effect at all.

Penrose went on to consider whether carriers could be prevented from breeding and to argue that such a scheme would also be impractical. Reiterating a point made in his 1935 paper, he noted that, though the gene may be rare, the carrier state is not. Assuming the incidence of the disease to be about 1 in 40,000, the Hardy-Weinberg theorem (which is used to calculate the frequency of heterozygote carriers from the frequency of the affected—that is, homozygous—persons) would show the carrier frequency to be roughly 1 in 100. (The incidence and carrier frequency are now considered to be about double what was assumed at the time.) Given those figures, a policy of targeting *all* carriers would be absurd. Even if they could be identified, eliminating "the gene from the racial stock would involve sterilizing 1% of the normal population," Penrose wrote. "Only a lunatic would advocate such a procedure to prevent the occurrence of a handful of harmless imbeciles."³⁰

But reducing the incidence of the disease seemed a worthy and practical medical goal. An unaffected sibling of a "phenylketonuric imbecile" has a 2 in 3 chance of being a carrier. Unless that person married a cousin or married into a community with an especially high prevalence of the disease, the spouse would have the ordinary 1 in 100 chance of being a carrier (given the then commonly assumed incidence of PKU in the United Kingdom of 1 in 50,000), and the probability that their child would be affected would thus be 1 in 600—too small a risk to justify discouraging such unions.³¹ But the risk is much higher in consanguineous marriages. Here Penrose drew a distinction between eugenic and medical aims, with the first oriented toward the distant and the latter toward the near future. In his perspective, reducing the incidence of PKU in the next generation was not a eugenic aim, even when achieved through the control of reproduction—that is, by preventing consanguineous matings in affected families and, if and when a functional test to detect carriers were developed, by preventing carriers from marrying each other.

Penrose's interest in methods to detect heterozygotes is rarely if ever remarked on, perhaps because it seems discordant with his reputation as an anti-eugenics crusader. The incongruity is a function of changes in the meaning attached to eugenics. In 1946, Penrose characterized efforts to prevent carriers from mating with each other as "dysgenic," since the net result would be a small increase in the frequency of the gene. At the time, such usage would not have raised eyebrows. Nor would talk of "eugenically acceptable" methods, which he also employed.³² But today, "eugenics" has extremely negative connotations, and despite recent efforts by several philosophers to rehabilitate the term, it is generally attached only to practices of which the commentator disapproves.³³ Thus, there are today several widely accepted programs to screen carriers, with the aim of reducing the incidence of particular genetic diseases. These include mandatory premarital carrier screening for β-thalassemia in the Republic of Cyprus; the *Dor Yeshorim* premarital carrier screening and matching program used by Orthodox Jews (who reject abortion) to avoid the births of infants with Tay-Sachs and other diseases assumed to be at high prevalence among Jews of Ashkenazi descent; and high school screening programs for both β-thalassemia and Tay-Sachs in Quebec and Australia. But these programs are typically characterized as "eugenics" only by their critics.[34]

A ONCE-OBSCURE DISEASE

Penrose publicized phenylketonuria and began the process of investing it with powerful symbolic meaning. Today, not only physicians but ordinary citizens know something about this rare disease, and the ability to intervene in its course is considered a major medical breakthrough. Følling's reputation soared with the perceived scientific and social significance of his discovery. Thus, an entry in a recent dictionary of medical terms notes that "Følling is by many considered the most important medical scientist not to receive the Nobel Prize for Physiology or medicine."[35]

But at the time, Følling's discovery attracted little attention. Until PKU was made the subject of Penrose's inaugural address, its existence was known only to a handful of researchers. In the 1930s and 1940s, the discovery of inborn errors of metabolism generated little interest, and although the condition identified by Følling was a real disease and not a more or less harmless biochemical variant, it was also rare and, at the time, untreatable. The obscurity of the disease is nicely illustrated by Penrose's account of a visit he made to an institution for the mentally retarded during a trip to the United States. After commenting that it took "a surprisingly long time for the significance of Følling's discovery to be appreciated," he wrote:

> In 1939, it [PKU] was unheard of in some of the grandest American clinics. When visiting a hospital in New Jersey in that year, I was shown a patient with a severe degree of mental retardation who had a private suite of her own. Everything was beauti-

fully appointed but there was a noticeable pervading smell of benzaldehyde and, perhaps also, of phenylacetaldehyde. I was informed that this patient was the daughter of a distinguished writer but that, in spite of obtaining all the best opinions in the United States, no cause for the defect had been found. Now I had learned from Sir Frederick Hopkins that when exposed to air, phenylpyruvic acid gradually changed to benzaldehyde. This is a rather insoluble substance and is difficult to wash out of clothes and sheets. The patient also had the fair hair and blue eyes which are commonly found in association with phenylketonuria. After examining the patient's reflexes I felt quite certain of the diagnosis and told my hosts what I thought. "Impossible," they said. "How can you come here and in a few minutes find something which all our best clinicians have missed?" Next morning, however, the wonderful dark green colour which appeared after adding ferric chloride to the urine, and which faded away again, confirmed my suspicions.[36]

The institution was Vineland, and the exceptionally well-cared-for patient was Carol Buck. But two decades would pass before Pearl Buck would finally learn the cause of her daughter's impairment.[37]

CHAPTER TWO

PKU as a Form of Cognitive Impairment

For scientists and clinicians in the early twentieth century, phenylketonuria was a scientific puzzle connecting biochemistry and symptoms; for affected children and their families, phenylketonuria meant living with a cognitive impairment. Before the discovery of an effective intervention to prevent neurological symptoms, individuals with PKU faced the same challenges and opportunities as anyone with a moderate to severe cognitive impairment. Generations of people affected by PKU would have depended on their families, neighbors, and the broader society to survive. Their level of support would have been determined by their specific impairment (cognitive level, seizures, behavioral symptoms) and by where and when they lived. The treatment of people with disabilities has varied through the ages with each community's understanding of the causes of impairment, the availability of remedies, and the collective sense of responsibility for others. Attitudes toward cognitive impairment influenced the treatment of people with PKU and affected how families responded to social pressures and practical challenges.

THE BIRTH OF THE TRAINING SCHOOL

Nineteenth-century institutions in Europe and the United States reflected the Enlightenment belief that human improvement was possible through systematic study and application of knowledge, rather than merely through prayer or other spiritual approaches. Unlike the undifferentiated poorhouse or general hospital before 1800, institutions in the nineteenth century increasingly focused on specific impairments: there were schools for the blind or the deaf, large public asylums for people with mental illness, specialty hospitals that treated specific conditions, reformatory schools for the poorly behaved, and training schools dedicated to people with cognitive impairment—such as the Vineland Training School for Backward and Feebleminded Children, where Carol Buck resided.[1] By 1850, there were training schools in Germany, England, and Switzerland. Most of these institutions were placed in rural settings: faith in the curative powers of a simple country life surrounded by fresh air and a clean environment reflected the ideas of philosophers such as Jean Jacques Rousseau, as well as the reality of increasingly crowded and dirty cities, which seemed to have obvious detrimental effects on physical as well as mental health.

There were also more specific cures for cognitive impairment. Working at the Salpêtrière asylum in France, Édouard Séguin reported substantial improvement produced by stimulating the motor and sensory systems to strengthen a weak nervous system. He argued that people with cognitive impairment had senses and ideas; the problem was that their behavior—the inappropriate actions that distinguished them as different—was not properly controlled by their will. His pedagogy used touch to awaken perceptual, cognitive, and moral capacities in a standard order. Séguin's methods started with physical exercises to awaken dormant senses, then progressed through objects and activities to stimulate each of the five senses. The final stage was a variation on the moral treatment developed by psychiatrists for people with mental illness: creating a friendly environment was insufficient; people with cognitive impairment needed a teacher who expected—even demanded—that they learn. Such intervention was labor intensive, and it brought demonstrable results. By the 1850s, Séguin's methods were adopted throughout Europe and the United States.[2]

In 1848, physician and reformer Samuel G. Howe opened the first training school for idiots in the United States. He was influenced by Séguin, who soon joined Howe in the United States to spread his educational methods to training schools throughout the country. Such schools were opened with the hope that Séguin's methods would help reduce the populations of prisons and almshouses, where idiots seemed to reside in abundance. In the 1840s and 1850s, reformers such as Howe and Dorothea Dix argued that idiots could learn to be productive members of society. Working as a Sunday school teacher in prisons, Dix was outraged at the treatment of incarcerated women and spent two years cataloguing the abuses of people with disabilities in institutions throughout the United States. With Howe's assistance, she advocated for state and federal governments to set aside resources for training schools. By the 1870s, nine schools served approximately fifteen hundred students in the United States, and as many as 25 percent of the students were discharged as "improved" to their families and communities, as productive members of society.[3] The Vineland school, opened in 1888, was considered especially progressive. It replaced the standard large dormitory model with small cottages in 1892, and later created one of the first facilities for research on cognitive impairment, the Psychological Research Laboratory.

As physicians who were paid a steady salary to focus on a specific set of conditions, training school directors emerged as one of the first specialty groups in modern medicine. In 1876, Séguin and several colleagues formed the Association of Medical Officers of American Institutions for Idiotic and Feeble-minded Persons, later known as the American Association on Mental Deficiency (and today, the American Association on Intellectual and Developmental Disabilities). The stated purpose was to study the causes of cognitive impairment, improve training and education, and foster the development of institutions. Almost from the moment they were opened, however, the training schools began to mix the roles of caring for people with cognitive impairments and controlling the social problem of what to do with people who showed different levels of cognitive ability. By the 1870s, training school directors claimed expertise in administrative as well as scientific matters: they were experts both in running institutions and in identifying causes and cures of cognitive impairment and behavioral disorders. Indeed, medical treatises at the time included complicated schemes for classifying cognitive impairment based on a physical examination, such as "cretinism" and "Mongoloidism." The focus on medical causes reveals a shift away from education, the main purpose for opening training schools in the 1850s.[4]

COGNITIVE IMPAIRMENT AS A SOCIAL PROBLEM

Despite the growth in the number of training schools in the late nineteenth century, such institutions never had the capacity to treat all the people with cognitive impairment living in the community. The majority of people with PKU probably lived with their families, and their experience shifted as families' understanding of their role changed. In the United States before 1850, people with cognitive and physical impairments were generally included in everyday life, including school and church. By the end of the century, parents were often seen as the cause of the problem, and their disabled children were considered a terrible burden. Magazine articles, political speeches, and fictional stories map this change over the course of the nineteenth century, and many factors were involved: the growth of scientific attempts to organize and rationalize the world; emergence of the "common school" movement, in which efficiency of education was more important than cultural and moral training; and the temperance movement and similar health movements that identified personal—and family—responsibility for ill health.[5]

Institutions also struggled with cognitive impairment. Many of the training schools begun with much hope in the 1850s foundered by the end of the century, in part because cognitive impairment often turned out to be a chronic condition resistant to simple cures. Graduates who returned to their communities did not fare well without substantial support, and institutions took on a more custodial role for those unable to live at home. The emphasis shifted from cure of children to long-term care of adults, with "able-bodied" inmates taking on the physical labor requirements of maintaining the institution. E. R. Johnstone, the superintendent of Vineland from 1900 to 1945, was a leading proponent of the colony system, in which large farms were established on land at a remove from the parent institution. Intended to provide produce both for the schools and for the market, the colonies and other attempts at self-contained communities promised economic self-sufficiency. However, the promise was rarely fulfilled: both costs and overcrowding increased, while buildings and grounds deteriorated. With longer-term residents and inconsistent economic support from state governments, the institutions acquired a reputation as warehouses for the least fortunate. Public scandals revealing the poor conditions of idiot asylums were a common theme every decade from the 1880s through the late twentieth century.

Outside the institutions, people with cognitive impairment became a much larger social problem in the late 1800s and early 1900s. US census data suggested an increasing number of affected citizens: from approximately 15,000 in 1850 to more than 150,000 in 1900.[6] Based on the results of US Army intelligence testing of recruits in World War I, some commentators claimed that one-third of the male population of the United States met the criteria for some level of cognitive impairment—a spectacular increase from the rate of 1 in 600 that Howe reported in his survey of Massachusetts in 1848. Some of the increase in overall numbers was related to general population growth, but the increase in prevalence (number of affected individuals per population) is better accounted for by a critical change in the definition of cognitive impairment. In the early nineteenth century, French physician Jean-Étienne Dominique Esquirol described two levels of cognitive impairment: "imbeciles," who "enjoy the use of the intellectual and affective faculties, but in less degree than the perfect man," and "idiots," who "hear, but do not understand; they see, but do not regard. Having no ideas, and thinking not, they have nothing to desire; therefore have no need of signs, nor of speech."[7] Both imbeciles and idiots were readily differentiated from typical people, if not by physical stigmata, then in their inability to perform the tasks of an increasingly complex industrial society. Training schools focused primarily on idiots.

The movement toward universal education in the late nineteenth century revealed a new class of people with cognitive impairment: students who seemed "normal" but had unusual difficulty with learning. Special education rapidly advanced in this era, as a response to the needs of slow learners, and scholars such as Alfred Binet in France created tests based on simple tasks that the average child could do at certain ages. He believed that early identification would allow early intervention and that special education could improve if not cure mild delays in cognitive development. In the United States, Binet's work was popularized by Henry H. Goddard, director of Vineland's Psychological Research Laboratory, who invented the term "moron" to describe children and adults who appeared normal but scored well below average on adapted versions of Binet's tests. At Stanford University, Louis Terman built on Binet's work to create the first standardized intelligence tests (today, the Stanford-Binet Intelligence Scales). Such IQ tests were applied to adults, and a surprising number scored well below the mean. Thus, with IQ tests, a new definition of cognitive impairment emerged: rather than experienced physicians using physical examination to diagnose idiots and imbeciles, psychologists and educational professionals would use standardized tests of cognitive ability to detect morons as well.[8]

"Feebleminded" was commonly used to describe different classes of cognitive impairment in the late 1800s; by the early twentieth century, "mentally defective" became more popular as the increasing number of people classified as feebleminded coincided with new interest in the moral implications of cognitive impairment. In the early 1800s, idiots and imbeciles were generally thought to be dependent through no fault of their own, and sympathy, education, and material support were easily justified. This was in contrast to people judged to be lazy, weak, or otherwise morally impaired; such citizens needed harsh encouragement to get back to work. By the late 1800s, however, cognitive impairment was understood as a root cause of moral impairment. Alcoholics, criminals, tramps, paupers, and prostitutes were frequently thought to have limited cognitive ability; indeed, feeblemindedness was the cause of their moral shortcomings, since they could not distinguish right from wrong or foresee the consequences of their acts. In this context, training schools took on a new role: institutionalization could both protect society from mental defectives and protect mental defectives from the dangers of being left to the mercy of a complex and sometimes mean-spirited society.

Cognitive ability was widely believed to be inherited. Goddard persuaded many colleagues, especially in the United States, that it was inherited as a simple Mendelian recessive. But the view that mental defect could be reduced through selective breeding did not depend on acceptance of the recessive theory of its inheritance, or even of Mendelism; it only required the belief that the trait was transmitted from parents to children in the hereditary material. Given the assumption that feeblemindedness was indeed inherited, segregation of affected individuals during their reproductive years, or their sterilization, came to be widely embraced as a solution to the linked problems of moral failure and the feebleminded. Finding a solution came to seem especially urgent because the incautious feebleminded were assumed to have many children, whereas those who possessed good judgment and showed self-restraint had few. Eugenicists thus embraced various strategies to encourage children from certain couples and discourage children from others. Eugenic policies did not necessarily imply moral culpability. Indeed, Goddard became famous in part for his 1912 eugenics treatise *The Kallikak Family: A Study in the Heredity of Feeble-Mindedness*, yet he also sought to exonerate the poor and criminally minded by revealing their inherited mental defects as the cause of their immoral behavior. Their misconduct was not their fault, since they did not understand that their actions were wrong.[9]

Eugenic precepts suggested a partial solution to the problem of how to limit the social impact of feeblemindedness. Even with the increasing number of institutions and inmates—as many as fifty thousand people living in US institutions for the feebleminded in 1923[10]—there would never be enough resources to sequester all feebleminded individuals. As costs rose, sterilization emerged as one way to decrease the future population of the feebleminded and, as such, was seen as a critical adjunct to preparing individuals to live with their families in the community. By 1917, 12 states had passed sterilization laws, and people classified with cognitive impairment were involuntarily sterilized through the 1960s.

Despite the widespread support for sterilization laws in the United States, there were vocal opponents: Penrose, for example, used the example of PKU to argue that a eugenic approach to the problem of cognitive impairment was both immoral and impractical. It is also important to note that the number of sterilizations was relatively small and that IQ tests were not the only criteria; there is some evidence that women in certain circumstances welcomed "involuntary sterilization" by the state as a form of birth control when other methods were unavailable.[11] Finally, a careful examination of how courts defined feeblemindedness in cases of guardianship or sterilization reveals that judges were more likely to be swayed by the testimony of neighbors and social workers than by examinations conducted by physicians and psychologists. While experts debated how to detect feeblemindedness, how a person functioned in his or her environment seemed to trump expert opinion in the 1930s and 1940s.[12]

FROM THE MORON TO THE MENTALLY RETARDED

In the 1920s, the appeal of supporting people with cognitive impairment in the community began to grow. One reason was practical: the cost of institutional placement was high, and with the number of morons seemingly in the millions, community living was the only option. Institutional superintendents and other experts began to argue for the ability of the feebleminded to reintegrate into society through community-based services. New research challenged previously accepted ideas about the association of mental defect with immoral behavior, and the permanency of low IQ scores seemed less certain. Researchers in Iowa, for example, found that children's IQs could be improved with appropriate education and that institutionalization harmed children's emotional and cognitive development.[13] The strong performance of soldiers with low IQs in World War II further challenged the functional significance of such tests, and more generally, the subnormal IQs of many functioning adults were reinterpreted as evidence of the limitations of IQ tests rather than a hidden defect in individuals.[14]

The emergence of family support groups for children with cognitive impairment was probably the single most important factor in changing US attitudes. Mothers' organizations have a history in the United States extending back to the 1830s, and the precursors of parent-teacher associations started in the early twentieth century. One of the first parent groups for families with a child with cognitive impairment was the Council for the Retarded Children of Cuyahoga County in Cleveland, Ohio. As the name implies, "mental retardation" emerged in this period as a progressive way to describe children who were slow in their development but capable of learning; the contrast with the permanence—and stigma—of "feeblemindedness" was deliberate. Several more parent groups emerged in the late 1930s; then the number grew rapidly after World War II. By 1950, there were 88 local parent groups with 19,300 members.[15] These groups offered classes for retarded children, recreational and social groups, counseling and guidance for parents, and a variety of institutional services and informational services, such as providing equipment and coordinating social services.[16]

These parent groups were generally unaware of each other. Sessions for parents at the national American Association on

Mental Deficiency meetings led to a dramatic meeting in 1950, when some local groups decided to form a national organization. In the words of one parent, "Imagine it! Practically every parent there thought his group was the pioneer. Most of us were strangers to each other—suspicious of everyone's motives and jealous of their progress."[17] The pioneering group met in Minneapolis in 1950 to organize the National Association of Parents and Friends of Retarded Children; its core objective was "to promote and stimulate needed research into causes, cure and prevention of mental retardation."[18] The following year, its constitution was ratified by 23 local groups. In 1952 the organization was nationally recognized, and it soon changed its name to the National Association for Retarded Children (NARC). (Reflecting shifting sensitivities about terminology, the organization underwent three more name changes, finally becoming The Arc of the United States, or The Arc, in 1992.)[19] By 1960, there were 681 local groups affiliated with the NARC, and sixty-two thousand members.[20]

The NARC looked to medical charities such as the March of Dimes as a model of organization, fundraising, and political power. Originally named the National Foundation for Infantile Paralysis, the March of Dimes was founded in the late 1930s by Franklin Delano Roosevelt. Although he never publicly revealed that polio had left him unable to walk, Roosevelt lent his enormous prestige to this private effort to confront polio, an infectious disease that came in epidemics and left thousands paralyzed, after what started as a routine illness. Roosevelt's close associate Basil O'Conner was the energetic president of the association for 30 years: he used radio and Hollywood stars, as well as the president's popularity, to build a national movement. O'Conner ensured grassroots support for the organization by a simple fundraising idea—children collecting dimes—and by supporting the development of local chapters. Much of the association's funds were used to provide ventilators to communities hit by polio epidemics, but a portion of the funds went to scientific research. The licensing of the Salk vaccine in 1955 and the nearly immediate disappearance of epidemic polio from the American experience confirmed the potential of private organizations, university researchers, and government action in using science to solve health problems.[21]

The NARC differed from the March of Dimes and other voluntary groups in part because of the stigma of mental retardation (MR). While having a family member with paralytic polio brought great hardship, families with retarded children faced the vestige of early twentieth-century ideas that linked cognitive impairment, race/ethnicity, and genetic degeneracy to social problems such as substance abuse, dependency, crime, sexual depravity, and juvenile delinquency. Local NARC groups therefore spent much of their energy in recasting cognitive impairment as something that could happen to any family, regardless of race/ethnicity or social class.[22] That typical families constituted the membership of local support groups surprised at least one mother, who described her trepidation as she prepared to go to her first parent meeting: "I silently shuddered as I thought that we were joining the ranks of freaks in the eyes of our community. What we found was, for me, an amazing sight. There was a group of nice, average, Americans talking in that animated manner which characterizes any meeting of a PTA, service club, or church gathering."[23]

Although the stigma of mental retardation had receded somewhat, after its peak in the 1910s, parents of children with MR confronted the enormous pressure on all families in the 1940s and 1950s to be "normal." Expert opinion and the general public agreed that happiness and fulfillment were closely linked to strict gender roles and family togetherness, which typically centered around children. Retarded children did not fit this picture of the perfect family, and many experts urged institutionalization in order to preserve the emotional well-being of everyone in the family.[24] The famous psychologist Erik Erikson and his wife, Joan, for example, followed such advice in 1944 when they institutionalized their infant with Down syndrome and told their other children that the infant had died. Entertainers Dale Evans (Rogers) and Roy Rogers hid the nature of their child's condition until after her death at the age of 2. In this context, Pearl Buck's *The*

Child Who Never Grew, published in 1950, was groundbreaking: a famous author revealing her personal struggles with a retarded child. In 1953, Evans followed Buck's example and published *Angel Unaware*, a description of how her daughter's disability brought meaning to her parents' intensely public lives as Hollywood stars. Written from the point of view of her daughter with Down syndrome, Evans's book provided a different perspective from Buck's: the two years with Robin were a blessing that gave divine purpose to the Rogers family. *Angel Unaware* was a best seller, outsold in 1953 only by the Standard Revised Bible and by Norman Vincent Peale's *The Power of Positive Thinking*.[25]

Both *The Child Who Never Grew* and *Angel Unaware* provoked thousands of supportive letters from families: they tapped into a vast emotional well of people's struggles to understand how to live with a child with MR. In a similar way, local parent groups grew rapidly in the 1940s and 1950s, in part because they helped meet the need of families to explore the emotional aspects of being an imperfect family. This included both families who institutionalized their child and those that did not or could not. Many families that chose institutionalization found support for their decisions in local parent groups—the child was still a valuable part of the family, most agreed, even if not living at home. Even as the NARC gained national recognition and collaborated with prominent professionals and politicians, the organization maintained a high degree of parental control in its mission and leadership.

Local parent groups did more than provide emotional support; they were also politically active. For their children living in state and private institutions, parents lobbied legislators to improve funding and called on administrators to improve services. Such advocacy led to increased access to speech therapy, recreational opportunities, and training in activities of daily living, such as toileting and eating. For the roughly two-thirds of families that could not afford an institution or chose to keep their child at home, parent groups pressured schools, shops, and playgrounds to open their doors to their children. Sheltered workshops and recreation centers for adults with cognitive impairments also emerged in the 1950s as part of an effort to enhance community participation. In the late 1950s, parent groups took up the cause of scientific prevention. Their role, both in expanding federal support of programs for the mentally retarded and in promoting scientific research into causes and cures, is at the core of the history of PKU.

CHAPTER THREE

Testing and Treating Newborns, 1950–1962

❖ ❖ ❖

By the 1950s, scientific medicine offered hope that the medical conditions causing mental retardation could be prevented or cured. The transformation of PKU from an obscure metabolic cause of MR to a new paradigm for the treatment of a genetic condition to prevent MR, however, required a series of conceptual, laboratory, clinical, and practical steps. Scientists and clinicians needed to agree that high levels of phenylalanine were the problem; they needed to create an affordable infant formula that controlled the intake of phenylalanine but did not exclude other essential nutrients; they needed to demonstrate that such a diet, if started early enough, would change the clinical course of PKU; and finally, they needed to devise a simple and inexpensive screening test to identify the very few infants among thousands of births who might benefit from a special diet. Although Robert Guthrie is most often associated with the success of early treatment of PKU, accomplishing each of these steps required contributions from researchers across the globe.

INITIAL RESISTANCE

In the 1930s, several experiments that treated children with PKU by manipulating diet had come to naught. In Britain, Lionel Penrose's strategy of excluding protein foods from the diet resulted in severe malnutrition, and an alternative strategy of creating a synthetic mixture of pure amino acids was deemed prohibitively expensive. In the United States, George Jervis and Richard Block had proposed removing excess phenylalanine from natural protein through a charcoal-filtering process, but the work was not funded and the idea soon faded from view. It would be a decade before such experiments reappeared, and the idea of treating PKU with a special diet was greeted with skepticism from an unexpected quarter: Lionel Penrose. By the late 1940s, Penrose seemed to have lost hope in a nutritional therapy for PKU. He was surely discouraged by the failure of his own and others' efforts to produce a workable diet. But he had apparently also lost confidence in the notion that the metabolic and cognitive defects in phenylketonuria were causally related—and hence in the strategy of reducing dietary phenylalanine to mitigate symptoms of the disease.

Such a claim is likely to come as a surprise to anyone familiar with the literature on Penrose and PKU, and especially his celebrated 1946 inaugural address at University College London, After all, the most oft-cited sentence in his inaugural address is, surely, "There may be methods of alleviating the condition, even though it is inborn, in a manner analogous to the way in which a child with club-foot may be helped to walk, or a child with congenital cataract enabled to see."[1] Given Penrose's own experiments with low-phenylalanine (low-phe) diets, the common assumption that he had nutritional therapy in mind when penning that passage is understandable. But that assumption overlooks the fact that the address includes no allusions to diet. The (failed) administration of massive doses of thiamine is the only specific treatment mentioned, and this is immediately followed by a warning that "consideration of the associated abnormalities, such as small size of the head, might tend to make us cautious about expecting anything like a cure."[2] Similarly, the first edition of *The Biology of Mental Defect*, published in 1949, discusses PKU in detail but includes no mention of Penrose's own experiments with or prospects for dietary therapy.[3]

In his writings of the late 1940s and early 1950s, Penrose consistently expressed uncertainty about the causal connection between the biochemical and mental abnormalities in PKU.[4] Moreover, members of both major British groups that were working in the postwar period to develop a dietary intervention recall that Penrose was actively skeptical of such efforts. Thus, biochemist Louis I. Woolf writes that "Penrose, the leading British expert on PKU and the genetics of mental retardation, held that a single gene caused both the mental retardation and the abnormal chemistry as two independent effects."[5] John Gerrard, a member of the first group to successfully prepare and administer a low-phe formula, similarly recalls that "at a meeting in Oxford, at which Penrose was present, the great man scoffed at the suggestion that modifying the diet could have any influence on the child's behavior—he was so sure that the mental retardation was only associated with the metabolic derangement, and was not caused by it."[6]

FIRST EXPERIMENTS WITH DIET

Given Penrose's authority, it is not surprising that there was widespread doubt that the cognitive impairment associated with the disease was caused by an excess of phenylalanine and hence that PKU might be effectively treated by a phe-restricted diet. That doubt would explain why, in reporting on their investigation of several cases of PKU in 1951, Louis Woolf and his colleague David Vulliamy felt bound to explicitly argue that either phenylalanine or one of its metabolites, circulating at high concentrations in the blood, probably depressed mental activity and that reducing blood phenylalanine levels might therefore restore normal cerebral function.[7]

Woolf and Vulliamy hypothesized that blood phenylalanine levels might be reduced either by increasing the rate of excretion of phenylalanine or by restricting its dietary intake in infancy to a minimum. It later occurred to Woolf that one could build a phe-restricted diet in an artificial way by taking a mixture of all the necessary amino acids except phenylalanine and adding the required amount of fat, carbohydrate, minerals, and vitamins. But as Penrose had earlier discovered, single amino acids were extremely expensive, so the cost of treating a child for even just a few years would have been astronomical. However, while wrestling with this problem, Woolf attended a Biochemical Society meeting where the (Richard) Block and (Diana) Bolling method for estimating the amino acid content of various proteins was discussed.[8] The Block and Bolling method involved growing bacteria in a medium with all necessary nutrients except the one amino acid of interest. The protein carrying the unknown amount of this amino acid was then added to the medium and the growth of the bacteria measured. To prepare the medium specifically for phenylalanine estimation, Block and Bolling took a protein hydrolysate and ran it through a column of activated charcoal. Charcoal attaches readily to phenylalanine (as well as to a few other amino acids, including tyrosine), so this amino acid could easily be separated from the rest of the hydrolysate.

Before taking up his position at the Hospital for Sick Children at Great Ormond Street in London, Woolf had worked as a research chemist at the drug company Allen & Hanbury, which made protein hydrolysates for oral or intravenous use as a predigested food for the severely starved in postwar Europe.[9] As a result, he knew that such hydrolysates were safe and could be manufactured easily and inexpensively. Woolf was thus exceptionally well prepared to see that the Block and Bolling method could provide a simple way of preparing a mixture of amino acids that were free from phenylalanine: It was only necessary to boil casein (a readily available and inexpensive milk protein) with acid to break it down into a mixture of peptides and free amino acids, combine the resulting hydrolysate with charcoal, and then restore a very small amount of the phenylalanine, along with normal amounts of the other amino acids that were removed by the charcoal.[10] The resulting product could then be blended with low-protein foods, suitable for babies, that also supplied carbohydrates, fats, and other essential nutrients. However, Woolf's proposal to his superiors at the Great Ormond Street Hospital that they try to treat a patient with such a food was coolly received. He was told that his job was to devise new diagnostic tests, not "crazy treatments for conditions that everyone knew were incurable."[11] Moreover, since Woolf was a biochemist, not a physician, he relied on cooperation from his medical colleagues, who were loath to try this untested therapy on any of their patients.

But in early 1950, Woolf received a visit from Horst Bickel, a German émigré physician who had previously spent time in Woolf's laboratory. Bickel had worked as an assistant at the University Children's Hospital Zurich, where he had learned from Guido Fanconi about the ferric chloride test for PKU. (Fanconi was an ardent advocate for the use of biochemistry in clinical medicine.) On moving to University Children's Hospital in Birmingham, England, Bickel suggested that the urine of all retarded patients be tested.[12] In the third patient tested, a 17-month-old girl of Irish descent, the test was positive. Sheila Jones had the classical physical signs of PKU, including fair hair, eczema, and a mousy odor. She was also profoundly mentally retarded, was unable to stand, banged her head, groaned and cried for hours at a time, and took no interest in food or her surroundings.

Like Pearl Buck and Borgny Egeland before her, Mary Jones was unwilling to accept doctors' assertions that nothing could be done for her daughter. A single mother who relied on welfare, Jones waited for Bickel every morning in front of his laboratory, refusing to accept his claim that there was no known treatment for her daughter's condition. Unable to evade her even by using side entrances, and worn down by her perseverance, Bickel was spurred to reflect on whether something might be done. He and Gerrard considered whether Sheila's condition might be treatable with a low-phe diet, but they had no idea how to go about preparing one.[13]

At that point, Bickel consulted Woolf. With no hope of having his proposed artificial diet tried at Great Ormond Street, Woolf provided Bickel with its full details. Bickel later confessed that the idea seemed crazy to him, but that nothing would be lost by trying it. With Evelyn Hickmans, head of the Birmingham hospital's Biochemical Department, Bickel prepared sufficient low-phe formula to treat Sheila. In her biography of Robert Guthrie, Jean Holt Koch describes the dirty and time-consuming work, which had to be done in a cold room lest the mixture spoil. "The charcoal got on everything. Professor Bickel's figure wrapped in layers of sweaters, topped by a charcoal-smudged lab coat, became a common sight at the hospital." While his family enjoyed Christmas Eve, he remained at work in his frigid lab.[14] But both researchers and mother were convinced that Sheila's behavior improved appreciably as a result of treatment. Within a few months, she was able to stand and to walk with assistance, had made eye contact for the first time, and was calmer and more engaged with her environment. The head banging and continuous crying stopped. Her hair darkened, her eyes brightened, and the eczema and mousy odor disappeared.

Bickel's colleagues, however, thought that the observers were deluding themselves and that any behavioral changes were due to the extra attention Sheila received. Bickel and coworkers thus decided that the only way to prove the changes were actually attributable to the dietary treatment was to add phenylalanine back to the formula—and that this had to be done surreptitiously, to avoid biasing the mother. When Sheila failed to appear for the next weekly appointment, the researchers went to her home and found the mother distraught. Her daughter's condition had begun to deteriorate within a day of the addition of phenylalanine to the formula, and after a week, all the gains had been lost. Sheila no longer smiled or made eye contact and could no longer walk. The researchers explained what they had done and reintroduced the low-phe diet. When Sheila had regained the lost ground, she was readmitted to the hospital and, this time with the mother's knowledge and permission, phenylalanine was again added to the formula and the behavioral changes filmed. The film was shown at a meeting in Oxford, where it was apparently met with incredulity. According to a letter written by Gerrard in the 1970s,

Penrose was present, and his sole comment was "the child was better dressed when on the restricted diet than when off it."[15] (This remarkable 1951 film can be viewed on the Web.)[16]

Mary Jones's reaction to being told that phenylalanine had been added to the casein hydrolysate without her knowledge is unknown. Neither contemporaneous reports nor later retrospectives of the experiment provide any indication that researchers were troubled by the deception involved—a reflection of the evolving standards for experimentation with human subjects rather than any indication of a special moral obtuseness on the part of the investigators. At the time, and in many clinics well into the 1950s and even 1960s, such actions would have been considered ethical.[17]

When the success became known, Woolf's superiors had a change of heart and suggested that he start such a program at the Great Ormond Street Hospital. Bickel and coworkers had studied only behavior and had made no quantitative measurements; the film was their only evidence of treatment efficacy. Thus, Woolf and his psychologist colleague Ruth Griffiths, who had specialized in measuring the developmental quotients of infants, aimed to measure any IQ changes. They found significant intellectual progress in all three patients they studied at Great Ormond Street, but also recognized that improvement would likely be maintained only if patients remained on the expensive diet, and they concluded "that if costly treatment is to be justified it must be started early. *This means that the urine of every baby or young child in whom there is the slightest suspicion of mental retardation must be tested for phenylpyruvic acid.*"[18] Two of the children in this study were 2 years and 8 months old, and the other was 5 years and 5 months old. Despite the significant improvement, their intelligence level remained low, since the children had deteriorated so much during their first year or two. But Woolf and coworkers eventually chanced upon a child with PKU who was only 17 days old. Obtaining a suitable formula for a newborn was difficult, but the dietician at Great Ormond Street ultimately succeeded in preparing a liquid formula that the infant could take instead of whole milk. The child did not deteriorate as would be expected of children with PKU in the first year of life. This success prompted the researchers to suggest that all children born in the United Kingdom should have their urine tested with ferric chloride shortly after birth, with treatment immediately initiated in the case of a positive result.[19]

Unbeknownst to the groups in Britain, Marvin Armstrong and Frank Tyler at the University of Utah were also engaged in dietary experiments in the 1950s. Using individual amino acids that had been donated by pharmaceutical companies, Armstrong and Tyler were the first to use a mixture of pure amino acids to treat PKU. All five children treated with the synthetic diet seemed to benefit, but to a very limited degree. Armstrong and Tyler noted that Charles Dent and coworkers had placed a 2-year-old child on a phe-restricted diet for six months and observed no change in the patient's condition.[20] Given wide differences in experimental outcomes, their conclusions were cautious: it seemed probable that phe-restricted diets would prove effective in treating PKU, at least in some cases, but much more experimental work would be required to prove it. In any case, a phe-restricted diet would have to be initiated very early, before irreversible damage to the central nervous system had occurred.[21] Prompted by the Utah group's results, Frederick Horner and Charles Streamer at the University of Colorado placed two 4-year-old children on a phe-restricted diet, and they reported considerable behavioral improvement. More significantly, a newborn sibling of one of the children was followed up from birth and placed on the experimental diet as soon as a positive result was obtained with the ferric chloride test, at 8 weeks of age. After seven months of treatment, the infant was apparently continuing to develop normally.[22]

The first retrospective statistical study assessing the benefits of dietary treatment would not appear until 1960.[23] But despite the sparse and inconsistent evidence, a consensus quickly emerged in the late 1950s that treatment with a phe-restricted diet was probably beneficial in PKU, at least when initiated in infancy. The problem of an affordable and readily available formula was solved when several pharmaceutical companies developed commercial formulas based on casein hydrolysates. In the United Kingdom, Allen & Hanbury introduced the first manufactured product, Cymogram, in 1953. Within a few years, it was followed by Albumaid (Scientific Hospital Supplies), Ketonil (Merck, Sharp & Dohme), and Lofenalac (Mead Johnson).[24] In 1958, Lofenalac was approved as a medical food by the US Food and Drug Administration, based on experience with just six patients.

FROM TESTING TO UNIVERSAL SCREENING

Nearly every researcher who tried dietary treatment of PKU noted that the key was to start the diet early, before significant neurological impairment became apparent. Most children with PKU in the 1950s were identified long after substantial developmental problems were evident. One exception was the testing of newborn siblings of children already known to have PKU, but most new cases occurred in families without any known history of PKU. A transition from experiment to widespread clinical practice thus required a practical means of identifying at-risk infants.

In 1934, Følling had accidentally discovered that a few drops of ferric chloride added to a urine sample would detect the disease that came to be called PKU. That ferric chloride test was employed to determine the incidence of the disease among residents of institutions for the mentally retarded—studies demonstrating that PKU was the cause of retardation in less than 1 percent of institutionalized patients.[25] Institutional testing also identified PKU as the cause of retardation in specific individuals, such as Carol Buck. And once it seemed that dietary intervention might be effective in preventing or alleviating symptoms of the disease, at least when detected in early infancy, the urine of newborn siblings of previously affected children also began to be tested.

Population-wide screening first became practical in 1957, when pediatrician Willard Centerwall at the School of Medicine of the College of Medical Evangelists in Los Angeles showed that Følling's ferric chloride test-tube test could be adapted for much wider use in clinics, maternity hospitals, and doctors' offices.

Centerwall, like many researchers in the field, was the parent of a child with MR. He recognized that one simple modification was to eliminate the process of acidifying the urine before the ferric chloride was added. He recommended using enough ferric chloride solution at sufficient strength to intensify the reaction and thus counter the basicity of the urine. More important, he realized that if the test were performed directly on a wet diaper, there was no need to obtain a urine sample and to transport it to a laboratory. By directly applying the proper solution on a wet diaper, the test could be extended to all newborns—far beyond the limited number of children who were already cognitively impaired or were siblings of persons known to have PKU.[26] In what came to be called the "wet-diaper" (or in Britain, "nappy") test, a drop of ferric chloride solution was applied to a baby's diaper; if the amount of phenylpyruvic acid were excessive, a green spot would immediately appear. The test was administered when the baby was 5 or 6 weeks old, to allow sufficient time for phenylalanine, which is not present at birth, to accumulate and be excreted in the urine as phenylpyruvic acid.[27] Another modification, beginning in 1958, was the use of Phenistix dipsticks—paper strips impregnated with a buffered ferric salt—which were pressed against the urine-soaked diaper.[28] That method was especially popular in Britain, where it was well suited for use by home health visitors, but it was also used by small clinics and physicians in private practice in the United States.

Accounts of the history of newborn screening for PKU almost invariably focus on the development of the bacterial inhibition assay (BIA) by microbiologist Robert Guthrie in 1962, which replaced ferric chloride urine testing. But the near-exclusive concentration on "Guthrie testing" has served to obscure the excitement generated by urine testing of diapers. Indeed, population-based newborn screening using the wet-diaper test was first developed in the United Kingdom and, by the early 1960s, had already been routinized in several jurisdictions both there and in the United States. When the editor of a volume of 1959 conference proceedings on PKU wrote that "the prevention of the mental retardation associated with phenylketonuria represents one of the great advances in medicine," he was referring to the impact of population-based *urine* testing for the disease.[29]

In the United Kingdom, Phenistix testing was rapidly adopted. The first detection program, in Cardiff, Wales, began in March 1958. The program initially utilized the test-tube method of analyzing a urine specimen, but most mothers were unable to collect urine at home, and only about a quarter of the infants were being tested. The following year, the program therefore switched to the Phenistix method of testing diapers, increasing coverage to 95 percent. Screening was also instituted in Birmingham. In 1960, the British Medical Research Council convened a conference on the detection and management of PKU, which resulted in the Ministry of Health requesting Medical Officers of Health to consider routinely screening infants through the use of Phenistix. Although Local Health Authorities were not required to institute testing, within two years it had become routine in 131 of 145 authorities.[30]

In the United States, with its fragmented health care system, the uptake of urine testing for PKU was very uneven. It was most common in California, where 30 health departments began routine wet-diaper testing. There were also testing programs in Cincinnati, Ohio, and a few other communities. Although adoption was slower and patchier than in the United Kingdom, urine testing for PKU was of great interest to the NARC, to legislatures and government agencies, and to the media. According to a 1959 article in the *Saturday Evening Post*, "Today, thanks to a major medical break-through, PKU babies may grow up to be competent human beings, their intelligence unimpaired—provided the disease is diagnosed early enough, before the brain is damaged."[31] The US Children's Bureau (USCB) sponsored a multistate screening program with the aim of determining the prevalence of PKU, and in 1959, it recommended that hospitals, clinics, and individual practitioners routinely test for PKU, using ferric chloride on diapers.[32] Mandatory urine screening for PKU was advocated by some local units of the NARC and by a presidential advisory committee on mental retardation. Model PKU laws even circulated in several states.[33]

Enthusiasm for the wet-diaper test was sometimes informed by misperceptions of the magnitude of the problem of PKU. For example, the *New York Times* reported in 1957 that a new government program for preschool mentally retarded children emphasized early detection "because much retardation is affiliated to two hereditary diseases [phenylketonuria and galactosemia], both of which can be cured or improved by diet if caught in time."[34] But unlike the situation in Britain, with its system of home health visits, no effective mechanism existed in the United States for accessing the infant population.[35] Only 12 to 15 percent of infants were seen in the California clinics; the vast majority of babies, as elsewhere in the United States, received their care from private physicians.[36] Thus enthusiasm, whether well or ill conceived, could not translate into routine testing in the United States.

But there were also more general issues with urine testing. The urine had to be fresh, and such specimens were often hard to obtain. In the United States, new mothers often thought it inappropriate to bring in their infant with a wet diaper, and so would change it just before their appointment.[37] In well-baby clinics, nearly a third of the babies failed to produce a wet diaper during the time of the appointment, and as a result, many infants were missed.[38] The test also gave many false-positive results. Most seriously, phenylpyruvic acid begins to appear in the urine, at the earliest, a month after birth, but blood phenylalanine levels reach 20 to 30 times the normal level within about a week.[39] Thus the test was unreliable until the infant was at least 6 weeks old and had possibly suffered irreversible neurological impairment. Another critical issue was the lack of any simple way to monitor blood phenylalanine levels, and hence to regulate the diet. These problems would be solved by Robert Guthrie, a microbiologist and nonpracticing physician, whose second child, John, was diagnosed with MR at the age of 6.

THE DEVELOPMENT OF THE GUTHRIE TEST

At the time of his son John's diagnosis, Guthrie was a cancer researcher at the Sloan-Kettering Institute in New York, and he was motivated by his son's impairment to think about the scientific prevention of MR. In 1954, he joined the staff of the Roswell Park Memorial Institute in Buffalo to continue his cancer research, and he and his wife, Margaret, soon became active in the Erie County Chapter of the New York State Association for Retarded Children, a local chapter of the NARC. In connection with his roles as vice president and program chairman of the chapter, Guthrie met Robert Warner, who directed a center for mental retardation at the University of Buffalo. Warner mentioned that he wanted to treat two young sisters with PKU, but needed a method to measure blood phenylalanine during treatment so that he could know whether the diet was successful in lowering their blood phenylalanine levels. Guthrie immediately realized that a method he was already using to measure substances in the blood of cancer patients could be adapted to this purpose.

Guthrie's cancer work utilized a bacterial inhibition assay to screen for different antimetabolites in patients' blood. (Antimetabolites are chemicals that inhibit a normal metabolic process; because they can halt cell division, they are commonly used in cancer chemotherapy.) That work was based on the principle of "competitive inhibition," in which a substance that normally inhibits the growth of bacteria no longer does so when another substance, added in large quantities, "competes" with the inhibitor and overwhelms its capacity to prevent bacterial growth. Using the same bacteria and technique applied in the cancer screening, Guthrie and his technician, Ada Susi, developed a simple test for blood phenylalanine.[40]

In this assay, a well-known chemical antagonist of phenylalanine, beta-2-thienylalanine, is added to an agar plate, along with spores of *Bacillus subtilis*. The bacteria are normally capable of synthesizing their own phenylalanine, but the beta-2-thienylalanine inhibitor blocks a crucial metabolic step in this process and

Robert Guthrie, holding an agar plate.
Courtesy of the Museum of disABILITY History.

the bacteria cannot reproduce. When a filter paper disc impregnated with blood serum from an affected individual is placed on the medium, the phenylalanine, at high levels in the serum, diffuses into the medium and is available to the bacteria, overcoming their phenylalanine starvation and producing an easily seen zone of bacterial growth surrounding the disc.[41] The amount of phenylalanine in the blood serum is estimated by comparing the diameter of the growth zones with zones produced on the same agar plate from discs of serum to which known amounts of phenylalanine have been added. The small amount of blood serum required for the test could be obtained through a capillary blood draw, thus obviating the need to obtain blood from a vein, a process that is often difficult with small children.

Although this method worked well for monitoring blood phenylalanine, the children thus tested were already impaired. Knowing that the sooner dietary treatment began, the better the prognosis, Guthrie, now at the University of Buffalo Children's Hospital, became interested in the possibility of screening infants for PKU before damage occurred. Coincidentally, a niece of the Guthries was diagnosed with PKU. Margaret Doll was about 14 months old in 1958, when urine testing identified PKU as the source of her disabilities. It was too late for treatment to make a difference, at least for her cognitive impairment. According to Guthrie, "The impact of this event convinced me to work on a screening test for PKU."[42]

That test involved only slight modifications of the one used to monitor treatment. The monitoring procedure made use of blood serum, which would be impractical for routine screening.[43] But it turned out that whole blood also worked and, furthermore, that the blood could be dried, thus avoiding the inconvenience of using liquid blood and a pipette to measure the sample.[44] As a result, the collection procedure could be greatly simplified. Blood obtained from lancing the heel of an infant could be blotted on filter paper, the paper dried, and an ordinary paper punch then used to obtain a disc from this dried spot for the test.[45]

This was potentially a huge advance over urine testing. The BIA was simpler, cheaper, and far more sensitive than the ferric chloride test. It could potentially detect blood phenylalanine levels by the third day of life, while the infant was still in the hospital—greatly increasing coverage in the United States and detecting PKU before the onset of irreversible neurological impairment.[46] Important as was the invention of the BIA, in the long run, what mattered more was the choice of blood as the analyte. Unlike urine, blood is highly stable, making it possible for samples to be sent to central laboratories and processed in batches. The BIA is only a "semi-quantitative" test, with the diameter and

BLOOD SCREENING FOR PHENYLKETONURIA
Robert Guthrie and Ada Susi

1. After puncture the baby's heel is touched to the filter paper. The specimens are then mailed to the lab.
2. After autoclaving, a small disk is punched out of each blood spot.
3. The melted agar culture medium, containing the phenylalanine antagonist β-2-thienylalanine, is innoculated with Bacillus subtilis spores, poured into flat dishes and allowed to solidify.
4. The blood paper disks from as many as 100 patients are, in addition to 7 control disks, placed upon the agar surface in rows, and after overnight incubation, the results are observed.

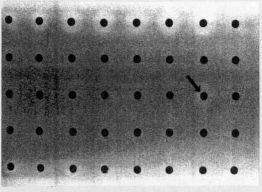

5. An indication of PKU is a blood disk producing a growth halo corresponding to any of the 4 highest control disks. This indication is then confirmed by a quantitative blood determination. The control disks are prepared from blood containing phenylalanine in the following concentrations: 2,4,6,8,10,12 and 20 milligram per 100 ml.

Robert Guthrie and Ada Susi's 1962 description of the bacterial inhibition assay for PKU. (Autoclaving the blood specimen was later found to be unnecessary.)

From Robert Guthrie and Ada Susi, "Trial of Phenylalanine Screening Method in 400,000 Infants (Guthrie Program)," Records of the US Children's Bureau, National Archives and Records Administration, College Park, MD, record group 102, box 950, file 20-12.

density of bacterial growth surrounding a disc generally proportional to the amount of amino acid in the blood. Other, more precise (although usually more expensive) quantitative ways to measure phenylalanine, such as paper and liquid chromatography, fluorometry, enzymatic analysis, and most recently, tandem mass spectrometry, would eventually compete with and partially supplant it. But the use of dried blood spots punched from filter paper has been a constant for the past fifty years. (Guthrie apparently acquired the idea of the filter-paper blood spot from a Sherlock Holmes story, where Holmes solved a murder by doing blood typing on a desk blotter.)[47] Moreover, because more blood was necessarily collected than was needed for the one PKU test, it would be easy to expand the system to include other tests.

Initially, it was unclear even to Guthrie whether the test would succeed in detecting most cases of PKU in infants. One question was whether blood phenylalanine increased quickly enough after birth to be detected in the first few days of life; another was the feasibility of large-scale testing. Thus Guthrie proposed both to the USCB and to Elizabeth Boggs, then president of the NARC and a member of President Kennedy's newly established advisory commission on mental retardation, that the BIA be field-tested to assess its suitability for a national screening program. Some metabolic researchers were unenthusiastic. The editor of the *Journal of Pediatrics* complained that "details of the method have not been published, hence there are no confirmatory studies. Nevertheless, the method has received wide publicity through the press, organizations of parents, and recently at scientific meetings," and suggested that more information on the validity of the test "should be made available before the Children's Bureau and state health departments embark on this mass testing program."[48]

Despite such reservations, in 1961, the Children's Bureau, with NARC support, commenced a large-scale field trial involving more than four hundred thousand infants in 29 states. In preparation for the trial, it was necessary to rapidly produce a large number of test kits. In consultation with Elizabeth Boggs and Gunnar Dybwad of the NARC, Guthrie decided it would

New Way to Detect a Dread Disease

Kammy and Sheila McGrath—the children featured in the NARC's 1961 annual poster campaign—with Lofenalac. Kammy (*left*) was 5 years old and living with her parents. Her sister, Sheila, was 7 and institutionalized. Because Sheila had been diagnosed with PKU, Kammy was given the "wet-diaper" test. As a result, she was treated with Lofenalac (heaped on table) and seemed to be developing normally.

From "New Way to Detect a Dread Disease," *Life Magazine*, Jan. 19, 1962, 45. Photographer Eric Schaal, © Time, Inc.

be most efficient to arrange for commercial manufacture of the kits. For this he favored the Ames Company, a subsidiary of Miles Laboratories, because it already marketed the ferric chloride urine test for PKU. However, company representatives said they would be interested only if a patent were issued, and in 1962, Guthrie filed a patent application in his own name and signed an exclusive licensing agreement with Miles. Under the terms of the agreement, he would not receive any royalties, but 5 percent of the net proceeds of sales would be divided among the NARC Research Fund, the Association for the Aid of Crippled Children, and the University of Buffalo Foundation.

When the company was unable to gear up quickly enough, USCB funds allowed Guthrie to rent space to establish a small "factory" for preparing a sufficient number of kits to screen a million infants, and training sessions were held in Buffalo for technicians from public health laboratories throughout the United States.[49] In the course of the trial, 39 cases of PKU were detected, indicating a surprisingly high incidence of about 1 in 10,000.[50] (The incidence had been assumed to be about 1 in 25,000.) By the time the trial ended, in 1963, the bureau had adopted the slogan "Test Every Newborn for PKU."

Boggs, a theoretical chemist with a PhD from the University of Cambridge, as well as the parent of a child with MR, was extremely enthusiastic about the new test. In 1961, the NARC decided to feature two siblings with PKU in their annual poster campaign. To publicize the field trial of universal newborn screening in conjunction with the poster campaign, Boggs urged Guthrie to publish a report quickly, and he (successfully) submitted it as a letter to the editor of the *Journal of the American Medical Association* in 1961.[51] Guthrie's first full manuscript describing the test was rejected by the *Journal of Pediatrics*, and a peer-reviewed report, coauthored with his chief technician, Ada Susi, was not published until 1963.[52]

Within a few years of Guthrie's publication, newborn screening for PKU became routine in the United States. The availability of a more reliable and cost-efficient screening test was a necessary but very far from sufficient condition for this development; cultural enthusiasm and political will and organization were also required to overcome substantial opposition. How screening for PKU quickly became a state responsibility is the subject of the next chapter.

CHAPTER FOUR

The Campaign for Mandatory Testing

Universal newborn screening programs are an anomaly in the United States, where medical practice has historically focused on the diagnosis and treatment of individual patients by their private physician, with the role of government generally restricted to regulation and payment. Although state hospitals and public health departments have long played a role in controlling infectious diseases, even most immunization programs have traditionally been implemented in the context of the private physician's office. Organized medicine has resisted attempts of government bodies to provide direct medical services to the general public, and government programs have focused on specific populations such as military veterans or the very poor.[1] The initiation of universal newborn screening (NBS) programs, in contrast, required that state governments invest in detection of specific medical conditions with relatively low incidence and no threat of contagion. What trends in medicine and social and government policy made such programs plausible?

THE TURN TO SCIENTIFIC PREVENTION

Preventing disease through diet or medicine was established as an ambition at least as long ago as 400 BCE, when Hippocrates began giving advice on personal habits such as diet and exercise.[2] Instruction on healthy living continued to be a common theme as modern medicine emerged in the eighteenth century. By the late nineteenth century, prevention of infant mortality was a central focus of medical and political leaders, with efforts focused on infant feeding and social conditions.[3] Specific medical interventions designed to prevent particular conditions became more common in the early twentieth century as physicians applied a scientific understanding of disease etiology. Leading advocates of preventive medicine urged universal application of such interventions as cod liver oil to prevent rickets, antibiotic ointment to prevent neonatal eye infections, and vitamin K to prevent hemorrhagic disease of the newborn.[4] By the mid-twentieth century, pediatric clinicians were convinced of the utility of universal treatment to prevent childhood diseases, and standard clinical protocols were common for newborns in US hospitals.

Recognition of the utility of systematic prevention is one thread that led to NBS programs; faith in scientific medicine was even more critical. The general public had been fascinated by the power of science since hearing popular stories of germ-fighting scientists and physicians of the late nineteenth century, but demonstration of the value of germ theory and other scientific discoveries was still lacking through the early twentieth century. In 1910, the infant mortality rate was still well over 100 deaths per 1,000 live births in most US cities, and nearly every family knew the tragedy of childhood death.[5] Highly publicized vaccination programs seemed to have little effect, and few new treatments provided the dramatic cures promised by the growth of laboratory science. By mid-century, however, public faith in the value of medical science was justified by the widespread use of penicillin in the 1940s and of the Salk and Sabin vaccines in the mid-1950s.[6] The polio vaccines, in particular, were critical to the public appreciation of the power of the laboratory to prevent disease and improve health. Many adults have vivid memories of polio, recalling how families lived in fear of summer epidemics when communities across the United States closed swimming pools and quarantined the ill, with the hope that what started as mild viral illness would not become a local epidemic of death and disability. By the late 1950s, polio had receded from the American landscape.[7]

While the polio vaccines confirmed the power of science to prevent disease, the identification and treatment of congenital syphilis solidified the idea of universal perinatal screening. As early as 1916, obstetrician J. Whitridge Williams required all women at his prenatal clinic at Johns Hopkins Hospital to receive a routine Wassermann test for syphilis, because, he asserted, early treatment could prevent syphilis in the infant. In the mid-1930s, public health officials noted that approximately sixty thousand infants were born with congenital syphilis in the United States. To prevent infant deaths and morbidity due to this disease, in 1938, legislatures in New York and Rhode Island issued regulations requiring blood tests in all pregnant women to prevent congenital transmission of syphilis. Similar laws across the country soon followed and seemed to be effective, even before the availability of penicillin: one California study demonstrated that the infant mortality rate for syphilis fell from 6.50 per 1,000 in 1938 to 0.15 per 1,000 in 1945.[8]

Among routine perinatal tasks, screening for and treating syphilis stood out because of the role of government in individual medical practice. The threat of epidemics of infectious diseases such as cholera led to an increasing role for local and state governments in public health through the 1800s.[9] In 1911, the high prevalence of syphilis, a reliable test for detecting *Treponema pallidum* (the causative bacterium), and the emergence of effective treatment led California to become the first state to require physicians to report cases of the disease.[10] In general, however, government did not become involved in the individual medical treatment of newborns. For example, vitamin K injections and applications of antibiotic ointment to the eyes were routine hospital practice but were not mandated or monitored by the state. In general, many newborn nursery practices were adopted only slowly by individual practitioners: although Rh factor was described in 1940, blood testing in newborns did not become routine for another three decades, even after the discovery of Rh immune globulin made it possible to prevent Rh blood disease—and significant cognitive impairment—in the newborn.

Despite a haphazard approach to adopting advances in medical practice, the general decline in childhood mortality in the United States seemed to confirm the value of scientific medicine. The so-called mortality transition describes the dramatic epidemiological shift in the United States and Europe from the early 1800s to the mid-twentieth century. In this almost two-hundred-year period, early childhood deaths due to infectious diseases decreased, while deaths from cancer and heart disease later in life increased. By 1960, for example, the infant mortality rate in the United States was less than 30 per 1,000 births, and childhood deaths were relatively rare.[11] The net effect of the mortality transition was an increase in average life span from less than 50 to nearly 80 years. Some authors have noted the relatively small role of antibiotics and vaccines in the decline of infant and child mortality, and debate continues on the contribution of improved nutrition, education, and sanitary practices in the mortality transition more generally.[12] But whether or not the credit was deserved, the decline was generally ascribed to advances in scientific medicine. Moreover, because families and physicians, by the mid-twentieth century, no longer faced daily deaths caused by infectious diseases, an investment of large-scale resources in relatively rare conditions seemed much more feasible.

At the same time, parents of children with mental retardation were mobilizing to increase public awareness of MR and to advocate both for improved services and for research into causes and cures. As noted in chapter 2, membership in the National Association for Retarded Children grew dramatically during the 1950s. At both the state and the national levels, the NARC and its affiliates were remarkably successful in lobbying Congress and the US Children's Bureau of the Department of Health, Education, and Welfare (HEW) to adopt policy changes to benefit people with mental retardation.[13] The condition first came to national legislative attention when Arthur Trudeau of the Rhode Island

Association for Retarded Children (a NARC state affiliate) presented his concerns to US Representative John Fogarty, a friend and former schoolmate, who chaired the subcommittee on health of the House Appropriations Committee. In 1955, Fogarty led a series of Congressional hearings, asking government officials what was being done on behalf of mentally retarded people. Martha Eliot, chief of the Children's Bureau, responded that MR was one of its three future major areas of focus. Substantial federal programs for people with MR, such as child development clinics, began in the wake of the Fogarty hearings. By 1959, HEW had devoted $40 million to such activities.[14] However, despite these efforts, the federal government as a whole remained only peripherally involved in this domain through the late 1950s, and some maternal-child health programs, by statute, were not allowed to serve children with MR.

Initially, the NARC had been oriented toward the expansion of educational, social, and rehabilitative services, an emphasis that coincided with the approach favored at HEW.[15] But Representative Fogarty thought that funds would be most productively used in a search for biological causes of mental illness.[16] And at the same time, the NARC's own position was shifting. In 1955, the organization hired university neurologist Richard L. Masland to lead a research effort into surveying current research facilities and the potential for identifying the causes of MR.[17] Masland, who became a member of the NARC's Scientific Advisory Board, wrote the report on prevention of MR, concluding, "The time is ripe for a large-scale attack on the genetically determined defects through the application of new techniques to the study of basic metabolism and to the study of patients suffering from mental deficiency."[18]

GOVERNMENT, SCIENCE, AND THE KENNEDY FAMILY

Federal involvement in mental retardation increased dramatically with the 1960 election of President John F. Kennedy, whose sister Rosemary was born with what would now be considered a mild cognitive impairment. Rosemary's father, Joseph Kennedy, Sr., who amassed the family fortune, became involved more broadly with the issue of MR largely through the advocacy of Cardinal Richard Cushing of Boston, a close family friend and advisor. In 1946, Kennedy and his wife, Rose, founded the Joseph P. Kennedy, Jr. Foundation, named after their eldest son, who was killed in World War II. In the 1950s, the Kennedy Foundation began to fund MR projects, although the personal reasons for the family's interest in the issue were not publicly disclosed at the time. Initially, the foundation's focus was on the provision of schools and other services for people with MR, but it was redirected in the late 1950s to research on causes and prevention.[19]

The focus on science was related to Joseph Kennedy's decision to turn over control of the foundation to his daughter Eunice and her husband, Sargent Shriver, in 1958. Eunice Kennedy Shriver was a remarkable advocate: she used the Kennedy name, their foundation, and the family's influence in the federal government to advance research, change attitudes, and commit resources to people with MR. So great was her influence on federal policy that the National Institute of Child Health and Human Development (NICHD), created in 1962 partly through her advocacy, was renamed in 2008 in her honor—the only NIH institute to be named after a person. Shriver's focus on scientific research and university programs, as opposed to schools and institutions for people with MR, was prompted in part by Robert E. Cooke, who, in the late 1950s, became one of her closest advisors. Chair of Pediatrics at the Johns Hopkins University, Cooke had a special interest in MR, partly because he had two daughters with cri du chat syndrome, a genetic condition that results in mental retardation. Cooke's research into the etiology of MR was supported by both the NARC and the Kennedy Foundation.

In 1961, Cooke participated in the making of a documentary film about the plight of people with MR and their families in Maryland. *The Dark Corner* told the story of the Cooke family's experiences after the parents decided to ignore ubiquitous advice to institutionalize their children, who could neither walk nor talk, and, instead, chose to care for them at home. The film was shown at the White House and may have influenced Eunice Shriver's decision to tell her own personal story in the *Saturday Evening Post*.[20] Her 1962 essay, "Hope for Retarded Children," revealed that she—and the president—had a sister with MR, something the Kennedy family had earlier been unwilling to acknowledge outside a very small circle.[21] Indeed, the family's success at keeping Rosemary's impairment secret is reflected in the fact that, as late as the 1960s, published accounts of the family consistently misidentified her condition. Thus, a 1960 book on the Kennedy family described Rosemary as a teacher in the Wisconsin school for retarded children, where she was actually a resident. A *New York Times* story, appearing the following year, quoted John F. Kennedy as saying that MR "should be brought out into the sunlight" but, ironically, attributed the family's interest in the subject to Rosemary's having suffered cerebral palsy in childhood and noted that "she is now doing some teaching in an institution in Illinois."[22] Shriver's short essay thus represented a dramatic break with family tradition. Adding to the confessional literature of Pearl S. Buck, Dale Evans, and others at the time, it confirmed that MR could affect any family, no matter how powerful. Shriver argued that there was hope for people with MR to learn and be employed, if we invested in science, educated professionals, and created a more accepting society.[23]

Perhaps Eunice Shriver's greatest achievement was in convincing her brother to make MR a federal priority. As a senator, John F. Kennedy had been uninvolved with the issue. In 1957, when a Senate subcommittee on which Kennedy served was considering the first MR bill, he left the hearings early, before the NARC representative was able to testify.[24] But following his election as president, Kennedy appointed Cooke to his health care transition team, apparently at the recommendation of Eunice Shriver.[25]

Shriver also urged the president to appoint a national committee to study the problem, and soon after taking office, Kennedy announced a new federal initiative to address the problem of MR.[26] He stressed that far more Americans—five million—suffered from MR than from diabetes, polio, and other diseases that received far more attention and funding. He also noted that facilities for care of the mentally retarded were overburdened. State institutions averaged 367 patients above their rated capacities, and there were only five hundred full-time physicians to care for the 160,000 patients in public institutions. Thus there was a huge imbalance between the scope of the problem of MR and the efforts to address it. On current educational and rehabilitative approaches, he asserted that "the central problem remains unsolved, for the causes and treatment of mental retardation are largely untouched." The president said he would strive to double the amount spent by the National Institutes of Health in this field. He also announced the appointment of the President's Panel on Mental Retardation (PPMR), headed by Leonard Mayo, executive director of the Association for the Aid of Crippled Children. The panel was charged with appraising the adequacy of existing programs and creating a plan for the federal government to conduct "a comprehensive and coordinated attack on the problem of mental retardation."[27]

The membership of the PPMR reflected an emerging nexus of university researchers, federal agencies, the NARC, and the Kennedys. Psychiatrist George Tarjan, the PPMR's vice president, also served on the NARC advisory board. Elizabeth M. Boggs, a theoretical chemist whose son had MR, was vice chair of the panel's Task Force on Law and Public Awareness and author of its report "The Causes of Mental Retardation." Boggs was also a founding member of the NARC, the association's first female president, and chair of its Governmental Affairs Committee for many years. A proponent of the scientific approach to MR, she was instrumental in guiding MR policy in the 1960s and 1970s. Cooke was another PPMR member with ties to the NARC and the Kennedy family, and Eunice Shriver herself was appointed "Consultant to the Panel." Shriver, in turn, was strongly influenced by Richard Masland's contribution to *Mental Subnormality*, the fruit of the NARC's three-year research survey.

Despite a consensus that MR could be prevented and that children should be the focus, the PPMR was not monolithic in

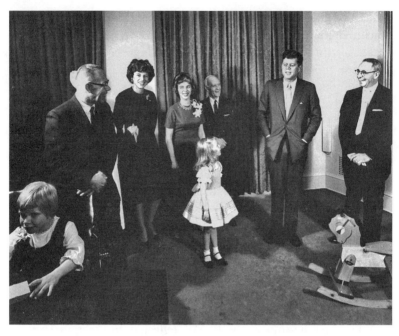

Kammy (*center*) and Sheila McGrath meeting with JFK and the President's Panel on Mental Retardation.
Abbie Rowe, White House Photographs. Courtesy of the John F. Kennedy Presidential Library and Museum, Boston, www.jfklibrary.org/Asset-Viewer/Archives/JFKWHP-AR6894-B-CR.aspx.

its approach. Among many points of contention, one critical divide was whether to invest primarily in basic research on biological causes or to focus on environmental deprivation and other social factors. The former tapped into America's broad faith in scientific medicine, as amply demonstrated by the success of the polio vaccine; the latter approach reflected an emerging political consensus around civil rights and the role of government in creating an equal playing field. Although the final report included a wide variety of recommendations for prevention and treatment, the thrust of the PPMR reflected the scientific orientation of the Kennedy administration and the NARC. Indeed, in announcing a major federal investment in biomedical research, the PPMR boldly predicted that application of medical science would lead to a 50 percent reduction in the prevalence of intellectual disability by the year 2000.[28]

PROMOTING UNIVERSAL NEWBORN SCREENING

In this context, Robert Guthrie's invention of a better screening test for PKU was hailed as a major medical breakthrough. The American Medical Association included the Guthrie test in its 1962 year-end report on major medical developments, along with the cracking of the genetic code.[29] In the same year, the Joseph P. Kennedy, Jr. Foundation presented an outstanding achievement award to Ivar Asbjørn Følling for his contribution to discovering PKU as a cause of MR.[30] At a gala for awardees, Følling was an honored guest at the White House, along with a representative of the NARC, members of the Supreme Court, prominent diplomats, and movie stars.[31] PKU was now a topic of discussion in leading newspapers and in magazines ranging from the *Reader's Digest*, *Parade*, *Good Housekeeping*, and *Life* to *Scientific American* and the *Saturday Review*. Both Følling and the disease he discovered had moved from near-obscurity into the limelight.

At the time, what chiefly captured public imagination was the ability to treat at least one cause of MR, which had hitherto been considered "the ultimate in therapeutic hopelessness."[32] Lamenting physicians' lack of interest in the field, Richard Masland conceded that the indifference was in part "a reflection of the undeniable fact that, in the majority of instances of mental retardation, medical treatment has little to offer."[33] In the context of a major federal initiative against MR and the rise of parent groups now focused on the search for cures, the Guthrie test appeared to be a major achievement, especially as other, similar successes were expected to follow. What made Guthrie testing "a breakthrough in the prevention of mental retardation" was both PKU's status as the first treatable form of MR and the expectation that it would not be the last.[34] From the start, Guthrie viewed his test as a template for the prevention of other diseases, especially other inborn errors of metabolism associated with MR. So did many others.

President Kennedy with recipients of the first annual Joseph P. Kennedy, Jr. International Awards for Research against Mental Retardation, posing at the White House. *Left to right,* Dr. Murray L. Barr, head of the Department of Microscopic Anatomy at the University of Ontario; Samuel A. Kirk, PhD, director of University of Illinois Research for Retarded Children; John Fittinger, president of the NARC; JFK; Dr. Ivar Asbjørn Følling, retired head of University Hospital Clinical Laboratory in Oslo; Dr. Jérôme Lejeune, director of the Department of Human Genetics at the University of Paris; and Joe Hin Tjio, an Indonesian geneticist.
Courtesy Bettmann/Corbis.

Duane Alexander, when director of the NICHD (1986–2009), recalled the excitement generated by Guthrie testing in the field of mental retardation, which had been "starved for success." He noted that "the optimistic view was that if we can succeed with PKU, then there must be other PKU-like conditions for which similar success could be achieved, if only we could do the research needed to discover them," and went on to comment, "This was one of the strongest arguments used by Cooke and Eunice Shriver

The Campaign for Mandatory Testing 65

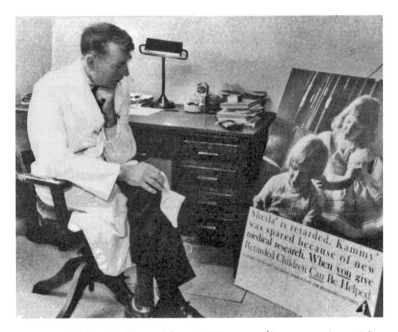

Dr. Ivar Asbjørn Følling with a NARC poster (in connection with the 1962 Joseph P. Kennedy, Jr. Foundation award).
Courtesy of the Joseph P. Kennedy, Jr. Foundation.

in championing establishment of a new institute at the National Institutes of Health to conduct, promote, and fund such research, efforts that culminated in the passage of legislation in October 1962 establishing the NICHD."[35]

Newspaper and magazine articles of the early 1960s were replete with claims about the importance of PKU as a model for the control of other disorders. According to *Parade* magazine, the most important value of the PKU test was "as a model for similar screening tests to uncover other unseen chemical and metabolic defects in the newborn."[36] A science reporter for the *New York Times* also viewed PKU as a template, noting that "the ailment is rare, but its importance is not to be measured in terms of numbers alone. It is one of those disorders that demonstrate the profound effects that a subtle, hardly detectable biochemical derangement can produce. The study of PKU promises to help uncover the causes of other hereditary disorders, particularly the basis for other types of mental deficiency."[37] After noting that fewer than five hundred individuals with PKU are born in the United States each year, another author wrote, "But the relatively low incidence of PKU is no measure of the attention it is currently receiving, especially from mental health experts. Their interest is easily explained: Never before has mental retardation been proved to have a known organic cause. The discovery of such a cause in the case of PKU has exciting implications. It suggests that, in time, certain other mental ailments—including schizophrenia and manic-depressive psychosis—may be found to have similar roots."[38]

Even before the 1961 Children's Bureau field trial of the Guthrie test had ended, several states had instituted screening programs for PKU. Robert MacCready, director of the Diagnostic Division of the Massachusetts Department of Public Health Laboratories and chair of the Public Health Services Committee of the NARC, had attended one of Guthrie's training sessions in Buffalo. (Like so many professionals who were proponents of screening, MacCready had a son who was cognitively impaired.)[39] Under MacCready's leadership, a voluntary program run by the state laboratory to test for PKU was established in Massachusetts in 1962. It not only was influential in demonstrating the feasibility of a mass screening program but also, in a stroke of good luck for screening advocates, identified three apparent cases of PKU in its first eight thousand tests, generating wide interest.[40] Given the low incidence of PKU, there could just as easily have been no detected cases; indeed, after finding nine individuals with PKU, MacCready's team did fifty thousand more tests without finding another positive.[41] As Kenneth Pass notes, that Guthrie first chose to direct his attention to PKU was fortuitous, for had he "chosen first to develop a test for homocystinuria (as he did later in his career), undoubtedly he would have been successful with that assay, but he probably would not have established newborn screening."[42]

At first, screening for PKU was simply added to the list of routine clinical procedures for newborns, such as giving vitamin K and applying antibiotic ointment to the eyes. The Massachusetts program, for example, received wide cooperation from physicians and hospitals in using this screening approach, reaching nearly 100 percent of births. However, MacCready argued that there was no excuse for missing a single child, and in 1963, Massachusetts became the first state to pass a law mandating screening for PKU. The next year, Rhode Island, Louisiana, and New York followed suit. Although several states achieved effectively universal coverage without legal mandates, by 1964, fewer than half of US hospitals had established screening programs, and overall, only about 20 percent of all newborns in the United States were being tested. Although the NARC had initially favored a voluntary approach, its members became impatient with the uneven uptake. Changing course after a year, its board of directors approved a resolution recommending the passage of legislation for mandatory screening for PKU.[43] It also circulated a model bill, which many states closely copied.[44] By 1965, 32 states had enacted screening laws, all but 5 making the test compulsory, and in the following two years, several other states did so as well. Thus, almost all the states that mandated screening did so within just four years, between 1963 and 1967. In some states, the laws passed with neither hearings nor floor debates, simply by acclamation or voice vote without dissent.[45]

By far the most important political player in state screening laws for PKU was the NARC and its affiliated state and local Associations for Retarded Children. "The NARC mothers were the beach commandos who landed on the state legislatures," according to one critical observer.[46] The NARC was well represented on the technical advisory committees that advised the Children's Bureau on mental retardation and newborn screening, starting in the 1950s; for example, two members of the NARC's Public Health Services Committee also served on the bureau's Technical Committee on Clinical Programs for Mentally Retarded Children.[47] The national organization had financially supported Guthrie's work and urged him to publish his report on the test quickly, so that it could be publicized in connection with the organization's 1961 poster campaign. Working closely with Guthrie, the NARC also used the media to directly influence state laws. In a typical example, one reporter asked readers, "How can you help to get this law passed? One way is to put pressure on your state legislature and public health departments. Another is to support organizations like the National Association for Retarded Children, which wants legislation requiring the use of all screening tests as they are developed and made available."[48]

The PPMR also played a key role in the passage of state laws. Even in 1961, when only urine testing was available, the newly constituted panel recommended that every state enact a law mandating PKU testing.[49] Its staff later urged the Advertising Council, which had been hired to publicize the magnitude of the problem of MR, to launch a campaign exhorting citizens in states without compulsory laws to demand them. The ad it produced featured a photo of an obviously healthy baby, with the caption "A fifty cent test spared this baby from spending his life mentally retarded." The text explained that only Massachusetts, New York, Louisiana, and Rhode Island legally required the test, and it advised, "If your state doesn't require the PKU test, urge its adoption. It should be a must for all babies everywhere." At about the same time, the national Parent Teacher Association also asked its members to urge their state legislators to enact laws making PKU testing of all expectant mothers and newborns compulsory.[50]

One theme of the campaign against MR launched by President Kennedy was the economic burden imposed by care, at public expense, for the severely affected, most of whom lived in residential institutions, and thus the financial benefit that could be expected from a successful effort at prevention.[51] Not surprisingly, the case for mandatory screening—whether expressed in newspapers and magazines, or in bulletins directed at various organizations' members, or in testimony in federal and state legislative hearings—almost invariably emphasized its potential to reduce public expenditures as well as alleviate human suffering. Although the value of screening to the child and parents was often said to be "incalculable," such claims typically appeared in tandem with calculations of benefits to taxpayers. The predictions of substantial savings were sometimes based on false assumptions about the prevalence of PKU. For example, Senator Joseph Montoya, an influential member of the Appropriations Committee, asserted that "many" of the 5½ million mentally retarded in the United States suffered from PKU and that their impairment "could have been prevented if detected in infancy. Most of the state training schools for the mentally retarded are overcrowded and have long waiting lists for admission."[52]

One skeptic perhaps had the senator in mind when he asserted that "legislators were sold this program as a beginning means of emptying the overflowing mental institutions."[53] Such a prospect would certainly have appealed to state lawmakers in the 1960s. More than five hundred thousand persons still lived in such facilities, and state governments were grappling with the enormous financial and human costs of custodial care. Most proponents of legislated screening, however, acknowledged that relatively few beds were occupied by patients with PKU; after all, a nationwide Children's Bureau census in 1962 had identified only 399 children with PKU admitted to programs for the mentally retarded during the preceding five years.[54] Rather, proponents argued that, given the cost to the state of prolonged institutional care, a screening program would save money even if only a single infant with PKU were identified.[55] Often, the 50-cent unit cost of the test would be compared with an assumed $100,000 (or more) cost of lifetime institutionalized care for an untreated person with PKU. In more sophisticated analyses, the expenses for laboratory testing and evaluation and five years of treatment with the special diet were contrasted with the assumed costs to the state of institutionalized care for a proportion of the affected infants and the costs of medical and hospital care for the non-institutionalized.[56] Sometimes a figure for the loss to the community of the person's productive capacity was also included.[57] Although contemporary cost-benefit analyses typically ignored some important factors—for example, that institutions' costs are mostly fixed and thus small reductions in numbers of patients are unlikely to produce equivalent cost reductions—the conviction that testing would ease the burden on state budgets was certainly a powerful incentive to enact PKU statutes.

Cost-benefit calculations of the 1960s also ignored the costs of contacting families and ensuring proper treatment when a child with PKU was identified by a NBS test. This analytical approach may have made sense when state laws were first passed, because many hospitals simply added the PKU test to the work of their clinical laboratories and reported results to the child's physician. However, Guthrie believed that regional labs with dedicated NBS personnel were critical to avoid disaster: the PKU test was deceptively simple, and false-negative results could arise from failure to make sure the bacterial strain was consistent. Furthermore, the low incidence of PKU meant that many hospital labs would never see a positive test, and it would be easy to miss the rare case. Even when the lab found a positive result, the clinician was likely to never have heard of PKU and to not understand when and how to implement the diet.[58]

Guthrie envisioned regional collaboratives or state laboratories like the one in Massachusetts as the best way to ensure high-quality laboratory work and appropriate clinical follow-up. Working with the NARC and the Children's Bureau in the 1961 field trial of the Guthrie test, Guthrie helped establish the path of state-mandated screening with the use of public health laboratories.[59] By funding 29 state maternal-child health agencies to oversee the research, the bureau established expertise and bureaucratic responsibility at the state government level. Throughout the 1960s and beyond, the Children's Bureau, its successor, the Maternal and Child Health Bureau, and other federal agencies funded clinical trials, convened academic conferences, and disseminated reports on newborn screening for PKU and other metabolic conditions. Federal funding of state maternal-child health bureaus continues to promote newborn screening as a state government function.[60]

By the 1970s, most NBS tests were conducted by regional collaboratives or state public health laboratories or, in a few parts of the country, by private laboratories tightly regulated by state agencies. Most state governments also took responsibility for notifying families of the results and ensuring that infants received proper treatment. This model for NBS was adopted, in part, for mundane reasons. Guthrie proved right about the inability of hospital labs to properly identify PKU positives, and missed cases led to costly lawsuits for hospitals accused of medical malpractice. Hospitals were eager to give the responsibility for NBS to states, and in time, the "PKU model" came to dominate in nearly every region: state laws mandated collection of a dried blood spot in the newborn period, and a state or regional laboratory ran the tests and ensured clinical follow-up.[61]

The adoption of state-mandated NBS in the 1960s is all the more remarkable because of the scientific uncertainty regarding many aspects of diagnosis and treatment of PKU. Indeed, in the United States, mandatory screening for PKU was strenuously criticized on both technical and political grounds. The skeptics' concerns and the controversy they engendered are described in the next chapter.

CHAPTER FIVE

Sources of Skepticism

SCIENTIFIC AND CLINICAL UNCERTAINTIES

In the 1960s, when states began mandating newborn screening for PKU, much about the diagnosis and treatment of the disease remained uncertain. For example, a report of a 1966 consensus conference on PKU, sponsored by the US Children's Bureau, criticized "legislation which has included public pressure for a specific therapy at a time when competent medical investigators are not in agreement on the criteria for diagnosis, natural course of the disease or optimal therapy or therapies."[1] The American Academy of Pediatrics (AAP) opposed mandatory screening, and its Committee on the Handicapped Child was especially critical. Its statement in 1964 noted that only "minimal data" existed on the efficacy of the PKU diet and ended with the following comment:

> The enthusiasts, however, see only the possibilities, without taking into account parents who are either unwilling or unable to maintain dietary treatment. Over-rigidity of dietary management has led to early death, presumably from insufficient protein intake or hypoglycemia. Over-hospitalization for too rigid control

has deprived the child of the normal stimulation and affection of home and family, thus preventing normal psychological maturation. Unrealistic claims for guaranteed normalcy . . . have led to frustration and discouragement on the part of both pediatricians and parents.[2]

The specific concerns expressed in the 1960s and 1970s centered on a series of closely linked scientific and clinical uncertainties, some troubling to proponents as well as to opponents of legislated screening. One of the most vocal critics was biochemist Samuel P. Bessman, who believed that scientists and clinicians were fundamentally wrong about the most basic aspects of PKU. Based on the experience with other diseases that had been mistakenly attributed to a single enzyme defect, Bessman predicted that phenylketonuria would turn out to have multiple forms with differing clinical consequences.[3] He stressed that phenylalanine was an *essential* amino acid, and he predicted that the diet (which he termed "deprivation therapy") would damage children misdiagnosed as having the disease. He speculated that many children diagnosed with and treated for PKU might have developed normally anyway and that phenylalanine was not the culprit in those with mental retardation.[4] Charging that the "frenetic haste for legislation . . . stems from our tremendous eagerness to get cases to study," Bessman claimed, "We made a legal vacuum cleaner to find these children. We justified it on several different bases, depending on the audience. To each other, we said that screening and therapy are experimental and must be tested. But I would like to see a single law in the text of which, or the preamble of which, it is said, 'This is a law to compel people to take part in a research project.'"[5]

Although both his style and the substance of his views made Bessman an outlier in the community of metabolic researchers, many of his peers were also troubled by the myriad unanswered questions. The issues that confronted PKU experts in the first two decades of screening—some of which are still being debated today—included who should be treated; the value of the Guthrie test; the harms, duration, and efficacy of dietary treatment; and even whether excess phenylalanine was the cause of MR.

Which Children Should Be Treated?

When screening began, no one knew what proportion of infants with persistently elevated phenylalanine levels were at risk for MR. Were all infants with persistently elevated levels at risk, as Guthrie argued, or only those whose levels were very high? There was no consensus on which infants should be placed on a phe-restricted diet, and disagreements over where to draw the line were often bitter. The confusion was partly due to ascertainment bias. Before the advent of mass screening, PKU cases came to attention as the result of clinical symptoms. It was thus generally assumed that all individuals with PKU suffered from MR. One group of metabolic researchers noted that "the occasional 'atypical' phenylketonuric person with normal or near-normal intelligence was considered to be sufficiently unusual as to be worthy of reporting separately."[6] Mass screening uncovered a surprisingly high frequency of elevated phenylalanine levels in the general population. But that finding could be interpreted to mean either that the incidence of PKU, understood as a serious disease, was considerably higher than had been thought *or* that there were clinically insignificant forms of PKU, which perhaps should not be called PKU at all—or to mean both. Complicating the situation still further, it soon became apparent that high blood phenylalanine levels could result from a variety of causes other than deficiency in phenylalanine hydroxylase, the enzyme usually associated with PKU.[7]

This problem of "variant forms" bedeviled early researchers. The normal concentration of blood phenylalanine was known to be 1 to 3 percent (or, in the terminology then current, 1 to 3 mg/100 ml), and most researchers agreed that persistently high concentrations were harmful. But there was considerable confusion regarding the significance of moderately elevated blood phenylalanine levels.[8] In the original field trial of the Guthrie test, any infant with a persistent level above 4 mg/100 ml was considered to have PKU, and Guthrie and Susi's 1963 report on the test recommended that any result of 6 mg/100 ml or more be considered positive.[9] But by the end of the decade, a rough consensus had developed that about half of those conventionally diagnosed with the disease did not require treatment. The term "benign hyperphenylalaninemia" came to distinguish those cases both from mild to moderate forms of PKU and "classical" or "severe" PKU, defined as a blood phenylalanine level of 20 mg/100 ml or more.[10]

The Reliability, Sensitivity, and Specificity of the Guthrie Test

Some researchers were unhappy with the Children's Bureau's decision to field-test Guthrie's bacterial inhibition assay for its suitability in a mass screening program before details of the method had been published. Moreover, several researchers complained that the data submitted to the bureau were inadequate for determining either the specificity or the sensitivity of the test and that, as a result, some infants with confirmed high phenylalanine levels at 2 to 3 weeks of age might not require dietary treatment. For example, consultants to the California State Department of Public Health agreed that, although the BIA was promising, "it requires further evaluation and our knowledge of PKU needs to be more complete before mass trials on the basis proposed by Dr. Guthrie would be justified." In their view, more intensive studies and a focus on high-risk populations would be a better approach than "deploying practically all available resources in a mass Guthrie Inhibition Assay screening procedure."[11]

It did not help reception of his test that Guthrie, a microbiologist, was an outsider to the human metabolic research community and that his peer-reviewed report on the test was not published until 1963. Furthermore, it was originally assumed that the results of Guthrie testing would be compared with later, more definitive tests. For the field trials, Guthrie himself suggested that tests be run both on the blood collected in the hospital and on urine-impregnated filter papers, which the mother would mail back to the laboratory when the infant was 2 to 3 weeks old—a method that he assumed would avoid frequent false-positive results. But for several reasons, the follow-up urine test soon proved unsatisfactory and, in reality, little systematic comparison ever occurred. Thus there was no direct comparison of the BIA method with other screening tests, no effort to define its reliability in the face of other compounds that might be present in newborn blood samples, no assessment of the test's reliability when administered very early in life, and no agreement on what level of blood phenylalanine was optimal. Consequently, there was no clear "gold standard" for the diagnosis of PKU that would allow an accurate estimate of the assay's sensitivity and specificity.[12] Indeed, the first systematic effort to assess the Guthrie test did not occur until 1974: about 10 percent of PKU cases were missed because the infants were not tested or because the test did not detect PKU, and 94.9 percent of positive screening tests were *not* confirmed as "classical" PKU on retesting.[13]

Guthrie had considered a high false-positive rate a small disadvantage in comparison with the benefit derived from early detection.[14] That view was shared by most researchers and public health officials. It reflected a common assumption—then and now—that the costs in time, money, and stress and the risk of unnecessary treatment are much less significant than the harm done by missing cases of the disease (thus sensitivity is more important than specificity). But others worried that infants could be physically harmed by overtreatment and by the financial, social, and psychological costs that adherence to an unnecessary or too-rigid diet would impose on parents and children.

Possible Treatment-Related Harms

Concerns about possible detrimental effects of treatment were related to the large number of false positives, the uncertainty about treating children with intermediate blood levels of phenylalanine, and problems in dietary management. Restricting an essential amino acid such as phenylalanine in a normal child could have devastating effects and was a real possibility, because initial screening results were not always followed by confirmatory testing.[15] But more worrying than treatment based on false-positive results was uncertainty about who should be treated and how restrictive treatment should be, as well as widespread difficulties in managing the diet when it was overseen by inexperienced physicians and anxious parents. In 1965, for example, Mary Efron noted that two infants diagnosed with PKU in the first weeks of the Massachusetts screening program had died after being placed on the restricted diet before referral to the Children's Hospital PKU clinic, and without any monitoring of their blood phenylalanine levels in the meantime.[16]

Even experts found management of the diet a difficult task, especially during the first six months of life, when the child was growing rapidly and required a rapid increase in phenylalanine. Although too much phenylalanine could be disastrous, it was understood that too little could also be damaging or even fatal.[17] A delicate balancing act was required by physicians and by parents managing the daily diet. The difficulties were compounded by uncertainty about the exact phenylalanine content of foods and by the child's refusal of the diet due to the unpleasant taste and odor of Lofenalac, the only formula then available in the United States. Moreover, Lofenalac was sometimes mistakenly considered a complete nutrition, and children not given an additional source of phenylalanine would presumably have suffered phenylalanine deficiency.[18] Perhaps most important, it was often assumed that blood phenylalanine had to be kept at the level considered normal for unaffected individuals.

As a result, there was much concern in the metabolic research community that poor management of the diet, and especially overly rigid phenylalanine restriction, would lead to severe malnutrition and even death.[19] Moreover, several cases of actual harm resulting from inappropriate treatment were widely reported in the literature, at conferences, and in personal communications.[20] In 1970, metabolic clinician-researcher William Hanley and his group in Toronto, disappointed with the progress of their patients treated for PKU and concerned by reports of permanent cognitive impairment resulting from malnutrition in the first six months of life, conducted the first systematic study of malnutrition in PKU. Among their 16 patients treated before the clinic relaxed its treatment guidelines, the researchers found 3 with severe malnutrition and evidence of lesser but prolonged malnutrition in most of the rest, and they raised the possibility that MR had been produced by malnutrition resulting from a too rigid dietary treatment.[21] Looking back, Hanley recalls that

> in the sixties when we started treating them, we were learning. We thought we should keep the blood levels normal, at 80–100 micromoles . . . So we realized in the sixties that it was okay to let that go a little higher and that would prevent the nutritional problems. And it wasn't until the late sixties that we got the dietician/nutritionists involved who taught us about nutrition and who took over management of these kids. We were doing it by the seat of our pants without nutritional help. And I think that happened a lot in various centers in North America and Europe. But that was the sixties.[22]

Even enthusiasts acknowledged that some children had suffered harm in the very early years of screening, but they thought that the causes had since been addressed. Thus, in an internal Children's Bureau memo in 1965, Ellen Kang noted that "we have come quite a long way from the days of initial treatment studies of PKU when dietary restrictions of phenylalanine were further aggravated by the unpalatability of the first products. Serious deprivation states were observed with outright Kwashiorkor-like states, hypoglycemia and intractable seizures." She went on to note that the availability of better-tasting products and greater awareness of the complications arising from severe phenylalanine restriction had since eliminated the problem of phenylalanine deficiency and that "deaths reported in the literature are largely attributable to complications of treatment with an unpalatable earlier product which caused diet refusal and vomiting and led to severe malnutrition."[23] It seems that death was a rare consequence of inappropriate treatment of infants diagnosed with PKU. Even

Bessman, who actively tried to identify such instances, claimed to know of only four deaths between 1961 and 1966, with the two Massachusetts infants the sole US cases.[24] But other complications of dietary therapy were clearly common.

The Duration of Treatment

Advocates often claimed that the diet had to be maintained only until the age of 5 or 6 years, when gross brain development was complete. Relating the story of a child first diagnosed and treated at the age of 4, the author of a popular magazine article wrote, "Had she come into this world three years later, she could have been put on the low-phenylalanine powders by the time she was six weeks old, before any brain damage had taken place. By the age of six, she would have been taken off the diet, for no further damage can occur once the brain is fully developed."[25] But other metabolic researchers noted that myelination continues at least through the teens and possibly longer, and some worried that PKU might turn out to be a chronic disease, with troubling implications for dietary compliance. There was little empirical evidence on the consequences of terminating the diet, and the few studies that existed had followed up children for only short periods of time and reported conflicting results. Transcripts of the United States Collaborative Study of Children Treated for Phenylketonuria (PKUCS), a major longitudinal study conducted between 1967 and 1983 at 19 centers across the United States, reveal intense and sometimes emotional discussions of the issue. A comment at a 1971 meeting captures their flavor: "When should we stop the diet in phenylketonuria? Protocol has, at least until recently, suggested six, and last year we had a little heated discussion in which people who don't believe the diet should be used at all argued with those who felt it should be discontinued at six, those who felt that it should be discontinued at eight, and one person threatened to take his jacks and go home if the diet were stopped."[26]

As late as 1980, a study of diet-discontinuation practices at PKU clinics found striking diversity.[27] Many clinicians aimed to terminate the dietary treatment as early as possible, given the difficulties of maintaining the low-phe diet, whereas others attempted to keep patients on the diet at least through adolescence. Moreover, given the products then available, many considered it impractical to try to keep school-age children on such a hard-to-manage regimen. But as children matured, evidence accumulated that discontinuation of the diet at school age was associated with cognitive decline.[28] Not everyone found the evidence compelling, however, and the question arose of whether the likelihood of such decline should count for more than the emotional and social impact of continuing treatment. Explaining why their group in London thought that the advantages of stopping the diet had received insufficient attention, Isabel Smith and Otto Wolff wrote, "The parents feel great relief when dietary restrictions are lifted and the protein substitute need no longer be given. In general, our observations suggest that continuation of the diet past mid-childhood leads to serious emotional disturbance in some patients and often achieves only limited control of phenylalanine levels because of frequent dietary lapses." The group's policy was to terminate the diet at age 8, "an acceptable compromise" between competing therapeutic goals.[29] But in 1978, a collaborative study by the London group and researchers in Heidelberg found significant decreases in mean IQ after the diet was discontinued, and other studies seemed to confirm that cognitive functioning declined and school problems worsened with increases in blood phenylalanine levels.[30] By the mid-1980s, when the PKUCS ended, most metabolic clinics in the United States, United Kingdom, and continental Europe were advising patients to remain on the diet at least through childhood.[31]

The Efficacy of Treatment

In legislative testimony and newspaper and magazine articles, it was often claimed that infants treated for PKU would develop normal intelligence. According to a 1960 article in the *Reader's Digest*, published even before the advent of the Guthrie test, "Now by a simple urine test PKU can be discovered, then brought under control, and the child assured an unimpaired intelligence."[32] But when mandatory screening began, no early-treated infants had yet reached an age when their adult cognitive functioning could be predicted. Metabolic researchers were well aware that, as the authors of a 1966 review article stated, "Not enough data are available to determine the long-range effect of treatment of phenylketonuria. The total time of treatment is not long enough to determine if treated individuals will be able to achieve an independent economic and social existence."[33] Although the researchers' own group had achieved excellent results with children treated before 3 months of age, they also noted that reports by other investigators on the follow-up of treated patients were disappointing, even when diagnosis occurred early and therapy was promptly started.[34] Participants at the 1966 NARC-sponsored conference similarly stressed how little was known about the results of treatment with a phe-restricted diet and the dependence of any interpretation of these results on PKU having been correctly diagnosed. The authors wrote, "The assumption is generally made that phenylalanine or one of its metabolites is toxic to the developing brain. As yet there is no evidence to support this statement," and they went on to note that animal studies were inconclusive and that no correlation had been found between levels of blood phenylalanine or its metabolites and the severity of brain damage.[35] Moreover, when screening began, all infants with persistently moderate blood phenylalanine levels (below 20 mg/100 ml) were treated. If half or more of these infants were not actually at risk, as many researchers suspected, their inclusion in studies would inflate the apparent value of treatment.[36]

The Cause of Mental Retardation

Bessman argued that no cause-and-effect relationship had been established between high blood phenylalanine levels and the mental retardation in PKU. In his view, the brain damage was as likely to be caused by a deficiency in tyrosine as by an excess of phenylalanine. Indeed, in 1968, he (unsuccessfully) attempted to treat an affected child with tyrosine supplementation alone.[37] A few years later, he proposed that the mental retardation in PKU occurred in utero as the result of an interaction between a heterozygous mother, whose capacity to convert phenylalanine to tyrosine is sharply reduced, and a homozygous fetus, completely unable to synthesize tyrosine. In this view, the mother's inability to compensate for the infant's lack of tyrosine results in impaired protein synthesis and consequent failure of brain development.[38] Bessman attributed the apparent benefit of phenylalanine restriction to a placebo effect—the extra attention received by a child with PKU.[39]

The efficacy of the phe-restricted diet could have been established with a small randomized clinical trial, in which only seven infants were not treated, but there was a widespread consensus that it would be unethical to withhold a treatment considered efficacious by a majority of researchers.[40] Bessman repeatedly argued that questions regarding both the efficacy and the potential harms of the diet would be difficult or impossible to answer once all infants were screened and those diagnosed with PKU were treated. In a typical passage, he wrote, "All of these questions will be answered by further experimentation, we are told. But if all children with excess of some amino acid in their blood are placed on deprivation therapy, we shall never know the natural history of any new disease or what the actual risk of mental retardation may be."[41] Members of the Children's Bureau technical advisory committee were also troubled by the possibility that universal screening would preclude resolution of scientific and clinical uncertainties in PKU. That concern is reflected in their deliberations on the introduction of screening for tyrosinemia (a much rarer metabolic disorder, one form of which is associated with severe liver disease), where committee members concluded that experiments demonstrating that therapy was efficacious should be conducted "before medico-legal problems, which have arisen in PKU, prevent an objective scientific evaluation of this metabolic disease also."[42]

As an alternative to such a trial, in 1967, the Children's Bureau funded the PKUCS, which originally followed up on 224 infants diagnosed with phenylketonuria as a result of NBS. The study represented a systematic effort to investigate the effectiveness of dietary treatment by treating all infants, but to varying degrees. The first phase of the study compared strict versus moderate restriction of phenylalanine; the second, continuation versus discontinuation of the diet (at age 6, participants were randomly assigned either to continue or to stop the diet). The PKUCS findings demonstrated that the diet was adequate for normal physical growth, could result in near-normal levels of intelligence, and should be maintained throughout childhood, and that the most important factor in predicting IQ was the age at which the low-phe diet began.[43]

OPPOSITION TO MANDATORY SCREENING

Despite the many scientific and clinical uncertainties in the 1960s, most experts in the field agreed with the fundamental premise that mental retardation in PKU was caused by an excess of phenylalanine in the postnatal period and that early detection leading to appropriate dietary management would prevent the neurological impairment of classical PKU. This does not mean that they agreed with mandatory, universal screening in the newborn period. A case in point is that of Mary L. Efron, pediatrician, neurologist, and analytical chemist at the Massachusetts General Hospital, who also served on the PKU advisory committee of the state's Department of Public Health. Efron was a screening advocate—she pioneered the use of paper chromatography to detect other disorders of amino acid metabolism, using the same dried blood sample collected for Guthrie testing—but she believed testing should be voluntary.[44] In a 1965 response to Bessman's charge that she was excessively sanguine about the outcome of therapy for PKU, Efron protested that she was indeed aware of many problems in treatment and added, "Unfortunately, along with compulsory testing has come much publicity and emotionalism, which has resulted in pressure to treat as phenylketonuric all infants with high or even intermediate levels of blood phenylalanine, without further study."[45] In late 1964, the American Medical Association (AMA) House of Delegates had voted against compulsory testing for PKU on the grounds that it "might set a precedent for compulsory testing of newborn infants in hospitals for many other conditions."[46] Efron agreed with the AMA position and expressed the hope that the organization would be "alert and vigorous" in opposing any further laws.[47]

Efron was influenced by her experience with the two infant deaths in the first weeks of the Massachusetts PKU screening program, as well as by her libertarian political views.[48] The AMA position was consistent with its long-standing opposition to state intervention in medical practice. Furthermore, both the AMA and state medical societies believed that NBS legislation was aimed at health care providers; that is, it was intended to give parents a right of legal action against those who failed to test. In every state but Massachusetts, medical societies lobbied against legislation. Indeed, ordinary physicians continued to be skeptical of screening laws well into the 1970s. A 1974 study of the knowledge and attitudes of 1,092 American pediatricians, obstetricians, and family physicians revealed that only 26 percent believed the benefits of screening for PKU outweighed the costs.[49]

The AMA and state medical societies represented the only organized opposition to compulsory screening. Most other opposition came from individual researchers worried that reports of harm could undermine support for screening. "Erroneous diagnosis and treatment with infants with the milder forms of hyperphenylalaninemia has resulted in deaths from phenylalanine deficiency and in much needless heartache for families having to manage an unpleasant and highly contrived diet," warned two clinician-researchers.[50] The authors of a 1966 review article proposed that legislators be made aware of the risks of treating children who did not need treatment, explaining that "it has been assumed that the chief problem lay in the detection of this rare disorder. The difficulties and hazards of treatment have not been emphasized. Death as a result of treatment of phenylketonuria has been a common occurrence. Growth failure, anemia and rachitic bone changes are also commonly reported in treated children and previously have been accepted as not too high a price to pay for normal mentality."[51]

Metabolic researchers who opposed mandatory screening were unorganized, however, and had no political presence. They generally complained to each other rather than to the press or to legislators. The major exception was Bessman, who aggressively communicated his concerns to the media and regularly testified as an expert witness on behalf of physicians sued for having failed to diagnose PKU. To the dismay of many of his colleagues, Bessman disparaged the low-phe diet in articles, interviews, and letters to the editor. He was clever, articulate, and witty, and even his worst enemies—he acquired many—usually acknowledged his powerful intellect and scientific achievements, including the invention of several important medical devices.[52] Bessman was a larger-than-life figure who dominated any discussion, and some of his peers were clearly cowed by his take-no-prisoners style of debate.

Bessman was not opposed to compulsion per se, and he strove to distinguish his position from that of the AMA, insisting that he was not motivated by ideological opposition to state intervention in medicine.[53] Rather, Bessman emphasized the uncertainties surrounding the etiology, diagnosis, and treatment of the disease, arguing that "we have a critical lack of information about almost every aspect of PKU. We do not know the cause of the retardation. We do not know if every individual with the genetic defect is in danger of retardation. We cannot be completely confident of our diagnostic tools. We do not know with certainty how to treat it."[54] Given his scientific standing and aggressive personality, Bessman's views could not be ignored. The felt need to respond to them was a major impetus for establishment in 1967 of the collaborative study of diet efficacy in PKU. In time, Bessman's claim that "many" individuals with high blood phenylalanine levels had normal intelligence would come to seem quixotic, while his theories of PKU pathogenesis and his conviction that the diet was biochemically ineffective diverged increasingly from mainstream thinking. By the late 1970s, Bessman had become something of a pariah in the field of metabolism.

But some of Bessman's concerns were shared by many, including proponents of mandatory screening. Indeed, although his views were generally considered extreme, some researchers appreciated Bessman's willingness to provide a critical voice at Children's Bureau meetings and other medical policy venues, and some publicly agreed with his opposition to state-mandated screening. At a press conference held in connection with a March of Dimes conference in 1967, for example, the respected pediatricians and PKU researchers David Yi-Yung Hsia and C. Charlton Mabry joined Bessman in calling for the abolition of compulsory testing programs.[55] Furthermore, many agreed that the attention to screening obscured the need to carefully follow up and treat identified infants. A statement by the AAP's Committee on Nutrition typifies this concern: "It seems anomalous that comparatively little has been done either to establish good treatment practices in hereditary metabolic disease or to mobilize scientific resources to ensure an optimistic outcome for therapeutic endeavors, while so much emphasis has been placed on detection."[56] This was one worry shared by the leadership of the NARC. Elizabeth Boggs, PPMR member and NARC president, favored legislated screening but thought that its value depended on what followed. That attitude explains the NARC's tepid reception of a bill, introduced in 1965 by senators Winston L. Prouty (R-VT) and Edward M. Kennedy (D-MA), that would have permitted the Surgeon General to aid states in financing screening for PKU and other inherited metabolic disorders.[57] There were several reasons for the NARC's unenthusiastic response, but as Boggs explained at the time in an internal memo, the first was "that the Kennedy-Prouty lays great emphasis on screening but almost none on follow-up. We give false hope if we screen but do not provide careful monitoring of each positive case." She went on to note that, despite the known possibility of false-positive results, some children had been placed on the restricted diet with no confirmatory testing and that the phenylalanine levels and protein intake of children with PKU required careful monitoring, if the diet were not to result in "serious malnutrition."[58]

Boggs's concerns went beyond the follow-up of children who screened positive for PKU to include the broader social causes of MR more generally. As a member of the PPMR—where proponents of biological research into the causes of MR wrangled with those who favored a focus on social causes and interventions—Boggs was a strong advocate of not only a scientific approach to prevention but also the development of comprehensive community services. She was concerned that an exclusive focus on scientific prevention would neglect many other children with MR.[59] Gunnar Dybwad, executive director of the NARC, made a similar point in a warning to the NARC membership during his 1965 farewell address:

> NARC will not be able to lay claim to being the national voluntary citizen organization speaking for the field of mental retardation if we continue likewise to overlook the social frontiers of mental retardation. The President's Panel Report . . . has certainly amply documented the extent to which mental retardation is related to socio-economic and cultural causes alongside of the Biological factors, to us much more familiar. Socio-economic factors and cultural deprivation may sound to you like fanciful theoretical formulations but they refer to things that happen in your own communities and it is certainly your responsibility to point them out to your fellow citizens and insist on appropriate civic action programs with just as much conviction as you put into your efforts to persuade state health departments to promote universal PKU testing programs.[60]

Joseph Cooper, a political scientist at Howard University, repeatedly raised the concern that screening programs would divert resources from more pressing needs. Cooper often collaborated with Bessman and similarly testified at legislative hearings, and in letters to the editor and articles published in popular magazines, in interviews, and in other forums actively disseminated his skeptical view of compulsory screening for PKU. Cooper also served as a member of the Committee for the Study of Inborn Errors of Metabolism, which issued the influential 1975 report *Genetic Screening: Programs, Principles, and Research*. Noting that PKU accounted for less than 1 percent of patients institutionalized for MR, Cooper asked, "What of the other 99%? Other problems of more widespread significance . . . are in need of social and legislative action. What are we doing, for example, about the home-situated retardees who awaken one day to find that their parents or relatives are gone or no longer able to care for them? What do we do about these people? They must certainly outnumber those with PKU."[61]

WHY THE CRITICS FAILED

Advocates responded to skeptics' concerns by arguing that the test was reliable, the diet was known to be efficacious, and the only way to resolve remaining questions was to identify, treat, and study children with PKU. Robert MacCready put the point succinctly: "Just as we go into the water to learn to swim, we must continue to search out, treat, and study the phenylketonurics."[62] Proponents also contended that screening did save money and that programs could be made even more efficient through regionalization of laboratories. Furthermore, increasing the number of tests to include other metabolic conditions would enhance the cost-effectiveness of NBS, since the major costs of screening were already incurred in collecting the dried blood spots and forwarding them to the laboratory. Finally, they reasserted that the importance of PKU lay in its role as a template for the treatment of other, more prevalent conditions. For example, responding to Cooper's claim that the emphasis on PKU would result in a slighting of the needs of the vast majority of those institutionalized for MR, Guthrie charged that Cooper "missed the point that those of us who have been excited about PKU realize that PKU is a 'model' . . . for future accomplishment with many of the other inborn errors of metabolism." He went on to note that "the specific prevention of MR in PKU has had a marked effect in increasing the interest of medical investigators and practitioners in the general problem of mental retardation."[63] Ultimately, proponents of mandatory screening for PKU argued that it would be immoral not to spare infants and their families the blight of mental retardation.

To establish state laws requiring testing for PKU, the NARC and its allies took their case directly to parents, the press, and the government. They lobbied legislators and published in local newspapers and in popular magazines such as *Life*, *Parade*, and the *Ladies Home Journal*. They were able to take advantage of the fact that, in the United States, screening was a responsibility of the individual states, where a handful of committed activists could be extremely effective. Screening advocates highlighted specific beneficiaries and victims such as the NARC poster children, Kammy McGrath, who had been diagnosed with PKU at birth and developed normally, and her older sister, Sheila, who had not and was impaired. Advocates understood and made skillful use of the human tendency to identify with named individuals far more than with featureless ones, an effect that is especially potent when infants and children are involved.[64]

Researchers, pediatricians, and public health officials who opposed mandatory screening for PKU disapproved of efforts to circumvent peer review. In their view, decisions about whether and what to screen should be made by experts and informed by scientific evidence, not driven by advocacy groups' lobbying and public relations campaigns. A common and related grievance was that newspaper and magazine writers routinely ignored the complexities and uncertainties of screening. The AAP expressed widespread frustrations in complaining that "new and complex information is often disseminated to the general public in an over-simplified manner by the mass media. As a result, there is often pressure to move with a sense of urgency when a new laboratory observation may have therapeutic benefits."[65] The AAP urged clinicians and researchers to resist such pressures.

But the concerns expressed by skeptics were abstract and often conjectural. Moreover, as Bessman noted, scientific controversy did not erupt until after the first three or four laws mandating screening had been passed, so momentum was with the proponents of such laws.[66] In the end, the skeptics had little impact on the establishment of mandatory screening programs or their trajectory. They were simply no match for the well-organized and politically savvy NARC and its parent volunteers and professional allies. State medical societies were more organized and had a history of success in the political arena, but their efforts in this case were strikingly ineffective. The concerns of clinicians carried little weight and were dismissed as self-serving when contrasted with appeals from suffering families, the prospect of saving money, and the opportunity to reduce the burden of state institutions that housed people with MR. As such, the state-mandated NBS programs constitute a small example of the more general creation of large government programs in the 1960s that were based on the promise of science and collective action, including Medicare, Medicaid, space exploration, and the war on poverty.

The skeptics in the United States did have a lasting impact on the separation of responsibilities for screening and treatment in state NBS programs. In responding to critiques by Bessman, the AMA, and state medical societies, Guthrie and his allies stressed that the state laws did not interfere in any way with the practice of medicine. Charging critics with confusing compulsory testing with compulsory medicine, advocates of legislated screening insisted that the state's only role was to ensure that testing occurred, with the child's physician responsible for deciding on follow-up tests and treatment. As Robert MacCready explained to the NARC membership, "The physician is still quite free to treat the patient as his enlightened medical wisdom dictates."[67] Although such a strategy had political advantages, it also reinforced a separation of laboratory screening from clinical care that had roots in the fragmented nature of the US health care system and the physical and institutional distance between laboratory and clinic. (In Canada and much of Europe, universal health insurance fostered greater coordination, as did the typical location of NBS in university medical centers.)[68] Looking back in 1978, Guthrie acknowledged that a legacy of the controversy of the 1960s was a lack of liaison between medical centers and even the best-organized screening programs in state health departments, with resulting problems in medical follow-up.[69]

By 1967, more than one million US infants had been screened for PKU, despite the great uncertainty in diagnosis, treatment, and long-term outcome. As a result, thousands of individuals who would have been severely impaired were enabled to lead normal lives. But several issues that perplexed researchers and clinicians in the 1960s continue to concern them today, and new issues have arisen. As we will see in chapters 7 and 8, these include the burden of dietary treatment for families and the devastating consequences of maternal PKU, a condition created by the success of screening.

CHAPTER SIX

New Paradigms for PKU

The remarkable improvement for individuals with phenylketonuria treated early in life led to many interpretations of the "PKU story." For human geneticists, screening and treatment represented a vindication of Lionel Penrose's prediction in the 1940s that the metabolism in PKU would someday be manipulated and, by extension, that other inherited disorders would turn out to be treatable.[1] In the words of Charles Scriver (who had studied with Penrose), treatment for PKU "changed the paradigm. Genetics is something you can't do anything about. Wrong. Genetics is something you can do something about and here's the example."[2] Through the 1960s, however, PKU programs also served as a paradigm in at least two other, at the time, more prominent ways: as a means to prevent mental retardation through the application of scientific medicine and as a novel public health program based on collecting, testing, and storing biological material obtained from an entire population. In the former case, PKU was presumed to be merely one of many metabolic disorders that would be discovered and treated to reduce the prevalence of MR; in the latter, universal newborn screening served as an unprecedented example of the role of the state in improving the health of individual citizens. Much like the example of polio vaccination campaigns in the 1950s, the experience with PKU in the 1960s solidified the US focus on laboratory medicine as the key to improving health.[3]

Over time, the "PKU paradigm" was increasingly identified with the geneticists' interpretation of the PKU story as proof that "genetic" ought not be equated with "fixed." As an exemplar of genetic malleability, PKU was increasingly invoked in the service of new agendas: in various iterations of the "nature-nurture" debate, it illustrated the flaws of genetic determinism, and in the discourse of biomedicine, it symbolized the value of genetic screening and the unique contribution of genetic research to the improvement of clinical outcomes. Indeed, wherever the point about plasticity is made, and to whatever ultimate purpose, it is likely to be illustrated by the successful alteration of the course of PKU.

Before PKU could serve as a model of effective intervention in a genetic disease, however, its being an *inherited* disease had to be emphasized to a much greater degree than had been the case in the 1960s, when most state screening programs were established.[4] Of course, it had been recognized since the 1930s that PKU was an autosomal recessive trait, but in the 1960s, the nature of PKU as a disease transmitted in the genes from one generation to another had little salience for the politicians, parent advocates, and others involved in the campaign to legislate screening for the disease.

Discussing the case of cystic fibrosis, Keith Wailoo and Stephen Pemberton note that in the 1960s and 1970s, "neither families nor experts emphasized the 'genetic' features of the disease. To be sure, they understood it to be a 'hereditary' disorder, but this way of thinking did not capture what they saw as its fundamental biological underpinnings."[5] Similarly, while popular articles on PKU testing almost always noted that the disease was hereditary and occasionally described its recessive mode of transmission, the focus was elsewhere: on the ability to prevent mental retardation resulting from at least one cause and on the possibility that a similar approach would prove effective against other disorders causing MR. That the devastating effects of the disease had far more salience than its etiology is reflected in the state legislative hearings on mandatory testing, where the hereditary nature of PKU received virtually no attention. None of the statutes establishing NBS programs even mentioned genetics. According to a 1975 National Academy of Sciences (NAS) report, interviews and transcripts of legislative hearings indicated that, at the time, most lobbyists and public officials involved in the process were responding to claims that screening would prevent MR and hence spare parents suffering and save taxpayers money. "There was little recognition of the implications for public policy, or for the impact on individuals who were screened, of the fact that PKU is a *genetic* disease."[6]

But the implications of PKU being a genetic disease would not have been obvious at the time most NBS programs were established. A majority of US states had launched screening programs by 1965, long before prenatal diagnosis for the disease existed.[7] The phenylalanine hydroxylase (PAH) gene was not cloned until 1983, and in any case, access to abortion was highly restricted in the 1960s. Guthrie testing did, incidentally, provide information on carrier status, since parents of affected infants would be obligate heterozygotes. But in the 1960s, the question of what difference it might make, if any, that a disease was genetic—and the associated issues of genetic discrimination, confidentiality, and privacy—did not yet appear on the policy agenda. They would emerge in the 1970s, following developments in both molecular and medical genetics.

GENETICS IN THE WIDER CULTURE: THE 1970S

The mid-1970s witnessed an explosion of controversy over the use and regulation of genetic technology. One catalyst was advances in molecular biology, which, even before the development of recombinant DNA (rDNA) techniques, prompted predictions of the "genetic engineering" of new genes and human qualities. Already in 1969, Cal Tech molecular biologist Robert Sinsheimer eagerly anticipated the emergence of a "new eugenics" that would overcome the limitations of the old variety. In his view, attempts to manipulate human breeding, a slow and clumsy process, would soon be replaced by direct genetic interventions—a prospect applauded by some commentators and deplored by others.[8]

Then in 1973, Herbert Boyer and Stanley Cohen spliced frog DNA into a plasmid of the bacterium *E. coli*, creating the first transgenic organism. This rDNA feat was followed a year later by Rudolf Jaenisch's creation of the first transgenic animal. The rapidly advancing ability to join molecules from diverse sources generated concern as well as excitement among elite molecular biologists, several of whom warned of the possibility that recombinant organisms of an unpredictable nature could be created and prove harmful to laboratory workers and, if they escaped laboratory containment, to the public. Their calls for strict procedures to prevent escape from containment and for a moratorium on very risky experiments culminated in the Asilomar conference of February 1975, which produced a temporary consensus that there should be extreme caution in rDNA experiments.[9]

Concern soon diffused beyond the boundaries of the molecular biology community. The US Congress held hearings on rDNA research, and some localities chose to implement their own regulations. Whether the new technology was dangerous to health or to the environment or would produce a dangerous (if voluntary) new eugenics became matters of intense public concern. With the rapid commercialization of the field (Boyer and Robert Swanson founded the first genetic engineering company in 1976), debates also erupted around the morality of "patenting life."[10]

Contemporaneous developments in medical genetics aroused a similar mix of enthusiasm and concern. Amniocentesis, the first practical method for detecting genetic disorders in utero, was developed in the 1960s, but it was of little practical utility before the 1973 Supreme Court decision in *Roe v. Wade* restricting states' ability to criminalize abortion. Following *Roe*, amniocentesis for the purpose of detecting Down syndrome quickly became a routine aspect of clinical practice in the United States. But the routinization of prenatal diagnosis also provoked controversy, especially around the question of whether policies designed to forestall the birth of affected children signaled a new eugenics.

Also in the 1970s, national and state legislation was first enacted to support research on genetic diseases, as well as to promote and regulate genetic screening programs. Responding to pressure from black professionals, celebrities, and community activists who argued that the incidence of sickle-cell anemia was much higher than that of diseases receiving far more attention and that the neglect was explained by racial discrimination, Congress passed the National Sickle-Cell Anemia Control Act in 1972, which provided funding for sickle-cell research, educational activities, and screening and counseling programs. In his signing statement, President Richard Nixon declared sickle-cell anemia to be an "especially pernicious disease because it strikes only blacks and no one else."[11] Four years later, Congress enacted the National Sickle Cell Anemia, Cooley's Anemia, Tay-Sachs, and Genetic Diseases Act, which permitted public funds to be used for voluntary genetic screening and counseling programs.

By the mid-1970s, many screening programs (under a variety of state, city, and nongovernmental auspices) had been established for sickle-cell disease and carrier status and also, at the community level, for Tay-Sachs disease. As there was no effective treatment for either disease, the primary aim of such screening was to provide reproductive information. But sickle-cell testing was soon engulfed in controversy when widespread confusion between sickle-cell trait (the generally benign carrier state) and sickle-cell anemia resulted in the stigmatization of carriers and sometimes discriminatory treatment in jobs and education.

In this context of heightened awareness of potential pitfalls in screening for genetic conditions, the question arose of what could be learned for the development of other screening programs from the relatively extensive experience of screening for PKU. The Committee on Inborn Errors of Metabolism of the NAS, chaired by distinguished pediatrician and geneticist Barton Childs, was charged with investigating the history, current standing, and effectiveness of screening for PKU and with reviewing screening programs for other genetic conditions such as the hemoglobinopathies (sickle-cell disease and trait and β-thalassemia) and Tay-Sachs disease.[12]

In its 1975 report, *Genetic Screening: Programs, Principles, and Research*, the committee concluded that screening for PKU was appropriate, and it rejected Samuel Bessman's provocative claim that many individuals with untreated PKU had normal IQs. But it was also harshly critical of the "rush" to enact PKU screening statutes in the mid-1960s, given unanswered questions about which infants with elevated phenylalanine levels needed to be treated and for how long, the efficacy of the low-phe diet, and the optimal level of blood phenylalanine. It described a political process in which legislators, hoping to save money and responding to massive pressure from local organizations of parents of children with MR, enacted statutes whose implications they did not fully understand. To avoid a repetition of what the committee characterized as "this experience of fragmented, uneducated, and hurried decision-making," it suggested establishing statewide or regional bodies to oversee the development of genetic screening programs and proposed a set of ethical, legal, and economic principles to govern their operation.[13]

The NAS committee considered newborn screening for PKU in the same category as any other genetic screening, which raised issues about the mandatory nature of NBS. Although informed consent was considered important in the domain of genetics, especially in reproductive decision making, the issue received little attention when most NBS programs were established in the 1960s. One reason is that the hereditary character of PKU initially received little emphasis, as noted above. Moreover, NBS developed under the auspices of public health, and public health has not traditionally incorporated consent into its programs. In the case of PKU, in particular, it seemed irrational for parents to refuse the opportunity to prevent MR in their child. Finally, the ethos of clinical medicine in the 1960s was still strongly paternalistic. Although neighborhood health centers and discussions of patient autonomy began in the 1960s, in general, physicians were expected to know what was best for their patients and to act on that knowledge. A legal right to informed consent for medical care was established in 1957, but through the 1970s it was common practice for physicians to make decisions for patients, particularly at the end of life.[14] The current practice of patient autonomy, in which well-informed individuals decide what medical care to accept, emerged slowly in the post–World War II period and became firmly established only in 1990, after the Supreme Court determined in the *Cruzan* case that people could refuse medical care even if lifesaving.[15] Thus, at the time that newborn screening for PKU was routinized, the question of whether to require explicit consent for testing was not on the agenda; the heel stick was simply added to the list of tasks performed by hospital personnel in the newborn period. While objections to mandatory screening had certainly been voiced in the 1960s, they were provoked not by concerns about consent but by providers' resistance to government intervention in medicine and by researchers' doubts that enough was known about the diagnosis and treatment of PKU to warrant compulsion.

By the mid-1970s, the desirability of informed consent for medical interventions, especially with respect to genetics, had become widely if not universally accepted. (Initially, sickle-cell testing was often mandated, but as opposition to the programs developed, many of these statutes were modified or repealed.)[16] The importance attached to consent in genetics is reflected in the 1976 federal law providing for assistance in the establishment and operation of genetic testing and counseling programs, which mandated that participation in such programs be "wholly voluntary."[17] And by then, PKU screening had come to be viewed as a genetic program. Thus, the lack of consent now appeared anomalous, and this feature of the program clearly made some members of the NAS committee uneasy. Although a majority apparently believed that compulsion in screening for PKU (unlike sickle-cell carrier status) was defensible under the "*parens patriae* doctrine that the state acts to protect those who cannot protect themselves" and that parents should not be able to refuse interventions clearly beneficial to their babies, the case for compulsion was argued half-heartedly and was hedged with provisos; according to the report, it *may* be justifiable *if* infants would otherwise not be tested and *if* detection leads to treatment.[18]

Whether reflecting weak or contested views or simply a recognition that the lack of consent was by then deeply entrenched (or a mix of ideological and practical considerations), the NAS committee report avoided strong recommendations. It noted that, although most PKU statutes provided for parental objections on religious or other grounds, there was generally no opportunity to exercise that prerogative—indeed, a mother might first learn of the existence of the test after the blood had been drawn—and concluded, "If mandatory laws are retained but such provisions are still to be taken seriously, changes will have to be made in the methods of taking specimens, so that parents are made aware of their right to refuse at an early enough time to exercise that right." The report also called attention to the increasing use of the blood samples taken to test for PKU for other metabolic tests and, in some states, the storage of innumerable dried blood spots "against the day when additional tests could be done." Noting that hardly any thought had been given to the issues raised by these practices, the committee presciently asserted that it was time to face the ethical and legal questions surrounding the storage of dried blood spots, "which tend to be regarded as a rich potential source for research purposes."[19] But the report did not prompt a reconsideration of consent procedures, and the issue largely faded from view.

It has reemerged only intermittently. By the late 1980s, it became possible to extract and amplify DNA from the huge number of stored Guthrie cards, which suddenly acquired potential research, diagnostic, and forensic value. This development generated discussion about the conditions under which consent should be required for the use of archived samples or for the future use of samples now being collected, but there was little impact on practice.[20] In the early 1990s, the Institute of Medicine of the NAS, responding to anxieties surrounding the Human Genome Project, appointed a committee to assess risks in genetic testing. Like the earlier NAS committee, it included NBS within its purview. The report concluded that while "it is appropriate to mandate the *offering* of established tests (e.g. phenylketonuria, hypothyroidism) where early diagnosis leads to improved treatable outcomes," it was inappropriate to mandate testing itself. In the committee's view, "no genetic test should be done without the consent of the person being tested or, in the case of newborns, the consent of the parents."[21]

The treatment of consent by both NAS committees reflects the anxiety about genetics that emerged in the 1970s. In the context of the rDNA and sickle-cell testing controversies, genetic information came to be seen as exceptional and in need of special protection. One common argument, based on the assumption that DNA constitutes our "essence," was that genetic tests disclose people's most basic characteristics. Another was that such tests uniquely foretell the future: genetic information was sometimes equated with a "future diary" that chronicles the most private aspects of a person's life. A third argument was that the results of such tests are particularly informative about relatives and thus raise distinctive issues of privacy and confidentiality. Whether or not these distinctions withstand close analysis, the doctrine of "genetic exceptionalism" became widely accepted and was deployed to argue for special protections for genetic information (as in state and federal statutes barring genetic but not other kinds of medical discrimination, such as the Genetic Information Nondiscrimination Act of 2008)[22] and to provide greater regulation of genetic than of other medical tests. Emphasizing the genetic etiology of PKU and adopting an inclusive definition of a genetic test as any test for a genetic disease potentially raised the ethical bar for the addition of new tests to states' NBS panels and could be used as a lever to argue for change in existing programs.[23] Calls for informed consent for NBS have increased as the number of conditions included has expanded to encompass diseases that are much more difficult to treat than PKU, but screening remains mandatory in nearly all states.[24]

A prominent theme in the 1970s was the need for caution in screening. Thus the 1975 NAS report stressed the complexities of diagnosis and treatment and described the history of screening for PKU as an "example of decision-making that was ethically questionable because of failure to consider enough facts."[25] A moral was that tests for other conditions detectable in newborns, as well as genetic tests more generally, should be assessed according to more stringent criteria than had been applied in that case. But in other contexts of genetic research and medicine, the PKU story was already taking a different and more consequential turn in which screening for PKU was stripped of all complications. The process of transforming a complicated into a simple story began in the 1970s with the controversy over the genetics of intelligence, and it intensified in the 1980s with the debates engendered by proposals to map and ultimately sequence all human genes.

PKU AND THE CRITIQUE OF GENETIC DETERMINISM

In 1969, Berkeley psychologist Arthur Jensen famously asked, "How much can we boost IQ and scholastic achievement?" His answer, in effect, was "not much." According to Jensen, genetic differences accounted for at least half of the black-white gap in IQ test scores, which explained why compensatory education schemes such as Head Start had apparently failed.[26] His essay, published in the *Harvard Educational Review*, produced a storm of controversy, with Jensen criticized both for exaggerating the significance of heritability estimates and for inappropriately generalizing from statistics on the heritability of IQ differences within races to conclusions about differences between them. (Heritability is the proportion of phenotypic variance in a population—such as the variance in IQ scores—that is attributable to genetic variance.) Two years later, writing in the *Atlantic*, Harvard psychologist Richard Herrnstein put forward an analogous argument with respect to social class, which he soon expanded to a book, *I.Q. in the Meritocracy*.[27]

The IQ debate had initially focused on the validity of genetic explanations for group differences in intellectual performance.

The heritability of *individual* differences was taken for granted. Studies published by British psychologist Cyril Burt had seemingly established a heritability of about 80 percent for IQ. But in 1972, Princeton psychologist Leon Kamin charged that Burt's results were, statistically speaking, too good to be true. After reviewing Burt's and the four other classic studies of the heritability of IQ, which he also found methodologically deficient, Kamin concluded that there were "no data which should lead a prudent man to accept the hypothesis that I.Q. test scores are in any degree heritable."[28] A heated debate followed on both the standards required to demonstrate the heritability of intelligence and the scientific and social value of heritability estimates.

Proponents of such research argued that it was both possible and desirable to design experiments on the heritability of human cognitive and personality traits, including intellectual performance, that met "reasonable" methodological criteria; that newer and better-designed studies had confirmed the existence of a substantial heritability of IQ, even if the new estimates were lower than Burt's; and that these results mattered for social policy. The message was typically that compensatory education and other policies designed to overcome the effects of poverty and racism rested on a naive belief in the power of the environment; to succeed, interventionist strategies needed to take genetic differences (as reflected in heritability estimates) into account. Critics of this research generally argued that heritability estimates of human mental and behavioral traits were scientifically and socially meaningless. They stressed the methodological difficulties involved in designing experiments that would break the association of genotype and phenotype (given that relatives generally share similar environments), and they insisted that the enormous efforts required to overcome this problem could not be justified by the potential scientific or social interest of the results. Heritability estimates, they argued, lack any policy relevance, since they are not a measure of the importance of genes in determining an individual's phenotype, are not generalizable (since heritability estimates vary with the mix of populations and environments), and above all, are not an index of plasticity.

In the context of this emotionally charged debate, the efficacy of treatment for PKU provided a dramatic, decisive, and easily understood rejoinder to the argument that a high heritability of IQ would render futile any efforts to boost scholastic performance. Critics stressed that PKU was a trait with a heritability of 1.0; that is, all the phenotypic variation among newborns is due to genetic variation. Yet an environmental intervention—limiting exposure to the damaging substrate (phenylalanine) through dietary changes—prevents otherwise severe neurological damage.[29] The ability to intervene in PKU thus demonstrates that a trait may have a high heritability and still be extremely sensitive to environmental change. Because the PKU case provided such a clear illustration that biology is not destiny, it came to serve as the standard illustration of the flaws of genetic determinism. In the early 1970s, critiques of Jensen and Herrnstein almost invariably invoked the ability to intervene in PKU, when arguing that research on the heritability of IQ was neither scientifically interesting nor socially meaningful. Indeed, the editors of a 1976 compendium of critical readings on the IQ debate wrote, "To use the standard example, consider phenylketonuria," in explaining how novel manipulations of the environment could not only improve the performance of a given population (in the PKU case, the IQ scores of treated children) but reduce individual differences as well (since the diet would raise the mean IQ of children with the disease but not of others).[30]

A few years later, a new debate erupted when Edward O. Wilson argued in *Sociobiology: The New Synthesis* that natural selection of inherited differences could explain such human social structures and behaviors as traditional sex roles, aggression, and altruism.[31] In this context, the case of PKU was again deployed to argue against genetic determinism. The following passage from a contemporaneous critique is typical: "There is an allele that, on a common genetic background, makes a critical difference to the development of the infant in the normal environments encountered by our species. Fortunately, we can modify the environments . . . and infants can grow to full health and physical vigor if they are kept on a diet that does not contain this amino acid. So it is true that there is a 'gene for PKU.' Happily, it is false that the developmental pattern associated with this gene in typical environments is unalterable by changing the environment."[32] And of course, the same example was invoked for the same purpose when the IQ debate reemerged with the 1994 publication of Richard Herrnstein and Charles Murray's *The Bell Curve*, which argued that intelligence is highly heritable and that differences in intelligence largely explained individual and group differences in social and economic status in the United States. Like Arthur Jensen, the authors also maintained that environmental interventions to raise IQ scores had proved largely futile and that to be successful, social policy needed to take facts about the genetics of intelligence into account. The debate resurrected all the arguments and counterarguments of the 1970s, including the same use of PKU. For example, one trenchant review of *The Bell Curve* argued that its authors were wrong to conclude that "equalizing environments will have no effect" on intellectual performance, for "it turns out that if you put all infants on a diet low in the amino acid phenylalanine, the disease disappears."[33]

PKU AND THE HUMAN GENOME PROJECT

While critics in the sociobiology and IQ debates cited PKU to argue that the social order is "not in our genes," others deployed PKU for a quite different and in some ways contradictory purpose: the defense of a genetic approach to medicine and, particularly, of the international effort to map and sequence the complete human genome. When the Human Genome Project was first proposed in the mid-1980s, many biologists expressed concern that it would siphon funds from other, more scientifically interesting efforts. Potential funders were also wary. Aiming to convince their peers, members of Congress, and the general public that this expensive project represented a worthwhile expenditure of public money, proponents mounted an intense public relations campaign that involved expansive claims for the scientific, technological, economic, social, and medical value of the project. The information gained would secure US leadership in biotechnology and promote economic competitiveness, produce technologies that would revolutionize many domains of biology, and generate "deep insights into the nature of humanity and our relationships to the world of which we are a part."[34] Above all, it would alleviate suffering. Thus, according to its supporters, the project would revolutionize medicine, resulting in cures for dread diseases—an aim with deep appeal to Congress. According to Harvard molecular biologist Walter Gilbert, a 1980 Nobel laureate in chemistry, co-inventor of a major technique for DNA sequencing, and cofounder of the biotech companies Biogen and Myriad Genetics, "The possession of a genetic map and the DNA sequence of a human being will transform medicine."[35] *Why* should it revolutionize medicine? The assumption was that locating disease-causing genes on chromosomes and determining their nucleotide sequence was requisite for a deep understanding of the causes of disease and hence for the development of truly effective interventions. As James D. Watson, the first director of the Human Genome Project, explained in summarizing his conversation with a congressman who had complained that he was tired of putting fingers in a dike, "In combating disease, genetics helps enormously if it is a bad gene that contributes to the cause. Ignoring genes is like trying to solve a murder without finding the murderer. All we have are victims. With time, if we find the genes for Alzheimer's disease and for manic depression, then less money will be wasted on research that goes nowhere." According to Watson, we thus need to convince members of Congress "that the best use for their money is DNA research."[36]

Critics were not impressed by such claims about the importance of genomic information for human health. They maintained that the project's advocates were overpromising, and they typically cited the "therapeutic gap"—the fact that genetic research had produced many more tests to diagnose or predict disease than means to effectively treat or prevent it, except by preventing births of affected individuals. They noted that "causal stories are lacking and therapies do not yet exist; nor is it clear, when actual cases are considered, how therapies will flow from a knowledge of DNA sequences."[37] They tended to be especially skeptical of promises that gene therapy, aimed at curing rather than mitigating the symptoms of disease, would follow from the possession of mapping and sequencing data.

In the context of controversy over the Human Genome Project, the PKU story acquired immense appeal to geneticists. With few effective interventions to point to other than abortion, the success in treating PKU came to function as a standard rejoinder to critics of the project and, more generally, as a way to legitimate both genetic research and the expansion of genetic testing. Already in the early 1980s, Henry Kirkman had remarked that neonatal screening programs for PKU "have been widely cited in textbooks of biology and genetics and in lectures to the general public," since they represented "one victory in the struggle against genetic factors, which are seen as being unalterable." And he noted that "PKU programs have become a showcase of the benefits to be derived from large-scale screening for genetic disorders."[38] With the controversy over the genome project, the case of PKU acquired even greater value to advocates of a genetic approach to medicine in general and the expansion of genetic testing in particular.

The legitimating role of PKU is nicely illustrated in a pair of National Public Radio programs in the 1990s. In the first, a caller to a *Talk of the Nation* show on the Human Genome Project asked about the relevance of genomics to breast cancer. He noted that the popular press was full of stories "about the magic of science and how genes are going to solve all of our problems," suggested that we should focus more on social context, and asked, "Has anything ever been solved by genetic research?" His question was tackled by Robert Waterston, head of the Department of Genetics and director of the Genome Sequencing Center at Washington University School of Medicine. "In terms of has it solved anything," Waterston replied, "there's a genetic disease called phenylketonuria—PKU—and just simply by testing infants at birth—and for those infants who test positive, if you give them a different kind of milk, you prevent brain damage. So this is clearly an instance."[39] Three years later, Francis Collins, then director of the US Human Genome Project, was interviewed on NPR. That interview was bookended by comments on PKU. At the beginning, after acknowledging that "the clinical consequences of genetics have been largely in the diagnostic arena up until now," Collins stressed that there are also genetic diseases for which treatments have been developed, "including the one that all newborns are screened for, the thing called PKU, where simply getting on the right diet prevents mental retardation." At the close, responding to a question about BRCA (breast cancer gene) testing, he replied, "I would say lives have been saved from this sort of genetic effort, actually, extending back some 20 or 30 years. And again, PKU is the example where the paradigm was proven."[40] Today, all types of genetic tests—for fetal anomalies, carrier status, and late-onset disorders—are still justified, at least in part, by reference to the success of newborn screening for PKU. As British geneticist Angus Clarke notes, that success has "accumulated a large store of goodwill and of ethical credit in favour of genetic screening programmes" generally.[41]

THE PRICE OF A PARADIGM

For PKU to legitimate the Human Genome Project, the success of screening had to be attributed to "genetic research" (in Waterston's words) or a "genetic effort" (in those of Collins). But screening and treatment for PKU were routinized in North America and much of Europe two decades before the PAH gene was identified and cloned. The ability to treat PKU rested on the biochemical insight that an excess of phenylalanine, which we obtain from the foods we eat, was somehow connected to the MR associated with the disease and that the symptoms might thus be mitigated if exposure to phenylalanine were reduced—an insight dating to the 1930s. The attribution of the ability to prevent MR to *genetic* research is a product of the 1980s and the controversy over the genome project. But it has now become standard, repeated even by those with no stake in promoting a genetic approach to medicine. For example, under the heading "Most Traits Are Affected by Environmental Factors As Well As by Genes," the authors of a well-respected genetics textbook write, "Here we come back to the low-phenylalanine diet. Children with PKU are not doomed to severe mental deficiency. Their capacities can be brought into the normal range by dietary treatment. PKU serves as an example of what motivates geneticists to try to discover the molecular basis of inherited disease. The hope is that knowing the molecular basis of the disease will eventually make it possible to develop methods for clinical intervention through diet, medication, or other treatments that will ameliorate the severity of the disease."[42]

Improved molecular understanding may well enable more effective or less burdensome clinical interventions in PKU. But to date, genetics has contributed remarkably little to either diagnosis or therapy for PKU. Both carrier testing for at-risk relatives and prenatal diagnosis for pregnancies at increased risk are possible if the specific disease-causing mutations in the family have already been identified, but the required analysis is complex and expensive, and neither procedure is widely used.[43] The cloning of the PAH gene was followed by enormous excitement about the prospect of gene therapy, but as with gene therapy more generally, those hopes were eventually disappointed.

Mutation analysis is sometimes helpful in predicting the severity of disease and in tailoring individual treatments. But there is great allelic variation in PKU, with more than five hundred identified mutations in the PAH gene, and most individuals with the disease are compound heterozygotes. This complexity has hampered efforts to develop diagnostic procedures based on genotype-phenotype correlation.[44] In treating patients, it may be useful to know whether a particular mutation is mild, moderate, or severe. In particular, genotyping may help predict which individuals will respond to a new therapy involving supplementation with sapropterin (trade name, Kuvan), an enzymatic cofactor for phenylalanine hydroxylase. But these are recent developments, and controversy remains about the value of genotyping individuals with PKU.[45] To claim that the prevention of brain damage in PKU is a clear instance of "genetic research" misstates the historical record.

The myth making has also come at a cost. As PKU acquired symbolic meaning, the story of its diagnosis and treatment became progressively simpler. Its cultural transformation into an uncomplicated success story began with the IQ and sociobiology controversies and intensified with the controversy over the Human Genome Project. Of course, the "PKU story" functions most effectively for these symbolic purposes when shorn of all complications. In the 1960s and 1970s, uncertainties, mistakes, and unintended consequences were widely acknowledged, at least among medical geneticists and professionals in public health. The 1975 NAS report stressed these complexities; in fact, it was precisely the difficulties of screening and treatment that made knowledge of the history of PKU potentially useful for the development of other screening programs. But as PKU acquired paradigmatic status, all nuance was lost. It became a disease easily and completely controlled by changing the environment, with no mention of the costs of screening and little attention to long-term adverse outcomes—bone disease, emotional disorders, and neuropsychological impairments—even in people who carefully follow the diet. Treatment is straightforward—in Collins's words, it is a matter of "simply getting on the right diet." But as anyone who struggles to adhere to the diet—or their family—knows, there is nothing simple about it.

The many stories of PKU reflect the needs of each generation and storyteller. In the 1960s, the PKU story confirmed the value of laboratory science in preventing MR and creating large-scale public health programs. In the latter part of the twentieth century, the story was deployed in a variety of contexts that reflected debates about genetics, even though its genetic nature had little clinical relevance. As we will see in the next chapter, the history of screening and treatment for PKU is best understood as the story of the transformation of a condition that causes profound cognitive impairment into a chronic disease requiring careful daily management. In this way, it perhaps most closely resembles type 1 diabetes, a disease transformed by insulin into a chronic condition that requires lifelong adherence to a taxing daily regimen and is associated with complications even when patients follow the regimen and are appropriately treated.[46]

CHAPTER SEVEN

Living with PKU

The management of phenylketonuria through nutritional therapy is inescapably arduous, given the sheer scale of the required reduction in dietary protein, the centrality of food to our cultural identities and social relationships, and the nature of the body's normal phenylalanine-control system, for which dietary therapy provides only an approximate substitute. We draw here on a series of in-depth interviews with individuals with PKU who attended a major metabolic clinic and on other firsthand accounts of life with PKU, to provide a glimpse into the challenges that the phenylalanine-restricted diet entails.

THE PKU DIET: AN OVERVIEW
"Diet for Life"

When mass screening for PKU began, it was widely assumed that only infants and young children required treatment. As discussed in chapter 5, in the 1960s, most experts believed that the diet could be safely discontinued at about 5 or 6 years of age, when gross brain development was complete. Moreover, given the unpalatable and hard-to-manage medical foods then available, it was generally considered impractical to keep school-age children on the diet. But guidelines later became more conservative, with recognition of the need for women to continue the diet at least through reproductive age (since phenylalanine is toxic to the fetus) and as studies found, first, an association of discontinuation of the diet with cognitive decline and, later, an association with more subtle deficits, such as mood disturbances and declines in attention span, memory, and ability to concentrate.[1] The age at which it was considered safe to discontinue the diet crept gradually upward. In time, most clinics in most countries adopted the principle of "diet for life," a policy endorsed by a consensus conference of the NIH in 2000 and reaffirmed in 2012.[2] Over time, too, target ranges for blood phenylalanine levels were generally lowered, and thus the diet became more restrictive.

Fifty years after the start of mass heel-stick screening, the low-phe diet remains the foundation of treatment for PKU.[3] In the interim, there have been major improvements in the nutritional quality, palatability, and portability of the dietary products. And for patients able to pay out of pocket or whose insurance helps cover the cost, low-protein (low-pro) substitutes for baking ingredients, pasta, rice, and an increasing array of premade food items such as crackers and cookies are now available. But the basis of therapy for PKU remains the same: avoidance of almost all regular sources of protein and consumption of what is variously (and confusingly) called a "protein substitute," "formula," or "medical food," which substitutes for intact protein.[4] The dietary regimen is onerous. The protein substitute, which should be consumed in equally divided doses throughout the day, is burdensome to prepare, transport, and keep cold, and it has a bitter taste and unpleasant odor. Even more significant, the foods that either are proscribed or are allowed only in small, measured quantities are ubiquitous. Indeed, the regimen is among the most rigorous medical diets, involving as it does both foods that must be ingested and foods that must be avoided—for a lifetime.

Frequent monitoring of blood phenylalanine levels is another feature of life with PKU. To ensure that levels are in the target range, they are monitored weekly or biweekly.[5] After infancy, the blood is obtained by lancing the fingertip, a procedure that many adults with PKU recall hating as children. Typical are the memories of a woman who recalls, "I was terrified of blood tests. My family would joke that they could hear me down the hall and out the front steps," and of a man who reports that, as a small child, "I'd have to be held, like my arm literally held down in order for them to draw blood."[6]

In one sense, remarkably little has changed since the 1960s, but in another, the landscape has been radically altered by the shift in treatment guidelines. The challenges of adhering to the PKU diet are greater for older children, adolescents, and adults than for infants and young children, who have never tasted foods that are proscribed and whose parents largely control what they eat. A Utah study found that blood phenylalanine exceeded recommended levels in only 9 percent of 2- to 4-year-olds, but the figure rose to 70 percent for individuals aged 5 to 18.[7] The difficulties are magnified for adults who were taken off the diet as children and learned later that they should resume it. The burdens of coping with such a restrictive, lifelong regimen explain why, as gene therapy researcher Cary Harding asserts and as interviews and comments on myriad PKU websites confirm, "A permanent cure that would eliminate sole dependence upon dietary therapy is the fervent dream of every individual with PKU, their families, and health care providers."[8] We turn now to the specifics of what it means to live with a disease that is treated with such a radical diet.

Protein Restriction: Avoiding Common Foods

Phenylalanine is one of the amino acids necessary for the synthesis of new proteins and for several other physiological functions.[9] Unlike plants and most microorganisms, mammals are unable to synthesize phenylalanine endogenously and so must obtain it from the protein foods they eat—it is an "essential" amino acid. The standard recommended dietary allowance for protein is currently 46 grams for women and 56 grams for men daily, with the mean actual intake in the United States considerably higher.[10] Since a gram of protein contains about 50 mg of phenylalanine, a typical American adult ingests about 3,500 to 5,100 mg of dietary phenylalanine each day. Adults with PKU are allowed approximately 350 to 500 mg, a daunting 90 percent reduction. Some are allowed even less.

Achieving a reduction in dietary phenylalanine of such magnitude requires not only excluding obviously high-protein foods from the diet but severely restricting the quantities of many other common foods that are not typically thought of as protein-rich. Meat, chicken, fish, eggs, nuts, soy, and dairy products such as milk, cheese, yogurt, and ice cream are obviously off the menu. Perhaps less obviously, wheat flour, rice, beans, lentils, pasta, potatoes, and even cauliflower, broccoli, avocados, corn, peas, and bananas are among many other ordinary foodstuffs that should be consumed only in measured, limited quantities. Indeed, the only foods wholly free of phenylalanine are water, refined sugar, pure starch, and pure oils.[11] Many people with PKU (and their friends, acquaintances, and sometimes even relatives) find it hard to believe that foods like pasta are *really* restricted. They know they should avoid meat, but as one young man put it, even though he was told that "beans were high in protein, it just didn't have the same bad ring to it." What is considered safe to eat and drink varies with the severity of the disease and thus the individual's tolerance for phenylalanine, but in general, regular pizza is considered off-limits, as are spaghetti with tomato sauce and sandwiches of any kind that are made with ordinary bread.

Also off-limits are tabletop packets of aspartame and foods, beverages, and drugs containing this artificial sweetener. On ingestion, aspartame breaks down into its components—methanol, aspartic acid, and phenylalanine—and the US Food and Drug Administration (FDA) requires that all products containing aspartame display the warning "Phenylketonurics: Contains Phenylalanine."[12] The ubiquitous labels have done much to publicize PKU, but the implication that *any* intake of aspartame is perilous has also generated unnecessary anxiety about its accidental ingestion, given that a 12-ounce can of diet Coke contains about 90 mg of phenylalanine—less than the amount found in a medium-size baked potato.[13]

A limited number of natural "free foods" that contain little protein do not count toward the daily total phenylalanine allowance and need not be measured. Metabolic clinicians differ in what they consider "free," as do national standards. The United Kingdom is quite liberal, including all fruits and vegetables (other than potatoes) that contain phenylalanine at less than 50 mg / 100 g.[14] Elsewhere, some clinics/clinicians interpret "free" to mean foods that contain only a little phenylalanine, while for others it means foods that are effectively phenylalanine-free.[15] The latter approach informs the *Low Protein Food List for PKU*, the bible for nutritional treatment in PKU. Many of the natural foods it identifies as "free" are actually condiments or ingredients in cooking or baking, such as Splenda, baking soda, cornstarch, rennet tablets, vegetable shortening, vinegar, sugar, honey, and salad dressings.[16] Parents searching for foods that a child can eat freely will not find many appealing choices. No wonder that many interviewees complained that they frequently felt hungry, especially as children.

On occasion, the PKU diet is analogized to a vegetarian one; indeed, to avoid convoluted and sometimes awkward explanations when they find themselves at a McDonalds or its equivalent and feel pressured to explain why they have ordered only a salad and fries for dinner, adolescents and adults sometimes say they are vegetarians. But the PKU diet is far more restrictive than a vegetarian or even a vegan one. After all, the mainstays of a vegan diet—tofu (a soy product), nuts, legumes, and grains—are prohibited or severely restricted for people with PKU. Even many vegetables and some fruits contain enough phenylalanine that the amount should be calculated and included in the day's total allowance. The numbers quickly add up. A banana contains 58 mg, and half of a medium grapefruit contains 49 mg. Some vegetables are strikingly high in phenylalanine: two *tablespoons* of peas contain 40 mg; a medium artichoke, 138 mg; a quarter-cup of cooked spinach, 164 mg. A single ear of corn on the cob would come close to using the entire daily phenylalanine allowance for most people with PKU. Even many standard vegetables contain significant amounts of phenylalanine; for example, a medium tomato contains 32 mg and a medium carrot 37 mg. So even at a salad bar one needs to be choosy.[17]

The magnitude of the necessary reduction in phenylalanine explains why it is "not unusual for someone who follows a low phe diet to have 2 kinds of vegetables and a baked potato for dinner."[18] A sample one-day food record published by the Mead Johnson company, which manufactures a protein substitute/formula called Phenyl-Free, illustrates the claim: dinner consists of a cup each of cauliflower and broccoli, a baked potato with margarine, and a glass of cranberry juice (as well as the Phenyl-Free).[19] What adhering to the diet means over the course of a day is illustrated by a *Teacher's Guide to PKU* plan for a 5-year-old boy.[20] His particular prescription calls for 13 scoops of Phenyl-Free mixed with 24 ounces of water and 280 mg of phenylalanine from ordinary foods. In the course of an average day, he might eat the following:

	mg phe
Breakfast	
Kix, 6 tbs	30
Banana, 6-inch section	30
Orange juice, 4 oz	15
Phenyl-Free, 6 oz	0
Lunch	
Vegetarian vegetable soup, ½ can	60
Saltine crackers, 2	30
Lettuce, shredded, ½ cup	15
French dressing, 2 tbs	free
Fruit cocktail, ¾ cup	15
Phenyl-Free, 6 oz	0

Snacks

Popsicle	free
Sucker, 1	free
Apple, 1 medium	8
Phenyl-Free, 6 oz	0

Dinner

Rice, cooked, 4 tbs	60
Green beans, cooked, 3 tbs	15
Jelly gelatin, 6 tbs	30
Phenyl-Free, 6 oz	0
Kool-Aid, 4 oz	free
Total	278 mg

The Protein Substitute

As a stand-alone strategy, restricting dietary protein will result in protein malnutrition and nutrient deficiencies, as well as a shortage of calories. For example, a typical 10-year-old child requires about 40 grams of protein and 1,800 to 2,000 calories daily; a child of the same age with PKU would be restricted to 5 grams of natural protein—the equivalent of about two slices of bread *or* a half-cup of milk.[21] Thus about 80 percent of protein needs and an even greater percentage of energy needs must be met by the protein substitute, which contains all of the necessary amino acids except phenylalanine, plus extra tyrosine, calories, and often vitamins and minerals. Although it is not uncommon for people who dislike or cannot afford the protein substitute to try to manage the disease with protein restriction alone, even modest reductions in protein intake are associated with a loss of energy, difficulties maintaining concentration, and other physical and neuropsychological problems.

The first commercial protein substitutes, such as Lofenalac in the United States and Cymogram in Britain, were powders that were dissolved in water. At a 1972 meeting, biochemist Louis Woolf commented on the difference in the composition of the formulas in Britain and the United States, which he attributed to the greater attention paid in Britain to the "social side-effects" of the PKU diet. According to Woolf, the British particularly valued protein substitutes such as Cymogram that were virtually phe-free and also low in calories, carbohydrate, and fat. "The aim in Britain," he explained, "was to keep the diet as near as possible and in as many respects as possible, to a normal diet for these children, and by that I mean normal in appearance, normal in texture, normal in taste, and as far as one could without raising the phenylalanine level too high, normal in composition. And to that effect, a lot of ingenuity and hard work went into producing special bread and special cookies which were free from phenylalanine . . . [T]his is really quite important from the standpoint of family stability and family structure."[22]

Because the protein substitute was intended for use by infants, it was called a "formula," and that term is still commonly used by metabolic professionals and by many people with PKU. However, given its infantile associations, some adolescents and adults dislike the term and refer to the product as their "milk," "shake," "juice," or "drink" or simply by its brand name.[23] To ensure that people receive enough protein, clinics typically advise consuming 24 to 32 ounces (0.7 to 0.95 liters) of the product daily, with the recommended amount depending on the individual's phenylalanine tolerance, on national standards, and on the policies of the particular metabolic center.[24] Optimally, the drink should be consumed in at least three servings, more or less evenly spaced throughout the day, to avoid wide swings in blood phenylalanine concentrations and to promote a sense of satiety.

In intact dietary proteins, the amino acids are bonded together; in the protein substitutes, individual amino acids are free. Unfortunately, free amino acids have notoriously poor organoleptic properties—that is, they taste and smell bad and feel unpleasant in the mouth. In the mid-1980s, Neil Buist, a metabolic researcher-clinician in Oregon, began to address this problem when he compared the amino acid profile of the leading medical food with that of cow's milk. He reported that the amino acid profile of cow's milk was based on acid hydrolysis, which distorts the results, rather than on the true profile as determined by protein sequencing. In addition, the current medical foods failed to optimize the minimum daily requirements for a normal child and contained large amounts of the most offensive (but inexpensive) amino acids. His group reformulated the mixture to improve both its nutritional adequacy and its palatability, eliminating the worst-tasting nonessential amino acids, reducing the amounts of several others (depending on their obtrusiveness), and providing more of the most acceptable. Soon, the profiles of all the commercial products were altered to improve their organoleptic as well as nutrient properties.[25]

In time, the protein substitute became available in a variety of flavors and forms. Recently, the number of choices has escalated. There are bars, capsules, and lipid-coated granules that are sprinkled on food. There is even an alternative food: glycomacropeptide (GMP), a whey protein by-product of cheese production that contains only small amounts of phenylalanine and lacks the amino acid aftertaste of the traditional medical foods.[26] As this is the first intact protein considered safe for people with PKU, efforts to turn GMP into palatable dietary products have generated enormous excitement. At least as important as improved palatability has been the development of more portable amino acid products. Individual packets of powder that can be added to cold drinks are now on offer, as are ready-to-drink juice boxes or "coolers," which are far more convenient to take to school or work and are particularly favored by college students.

Despite these newer choices, many people with PKU continue to use the traditional powder, which they mix with a liquid and take in containers to school, work, or other activities. The persistence of these powders is partly explained by cost considerations: the more palatable and easy-to-use alternatives also tend to be more expensive and are less likely to be covered by insurance. It is also difficult to get amino acids into the smaller forms, and as a result, the number of capsules that must be swallowed or bars consumed is forbidding. Instead of having to take more than 100 (or even 200) capsules or eat eight or nine bars daily, individuals

PhenylAde ads: "At Lunch" and "On the Go."
www.mixitin.com. Courtesy of the Applied Nutrition Corporation.

usually take these products to replace one serving of the liquid medical food, rather than eliminating it entirely. Moreover, the alternative products are fairly new, and many people have become accustomed to the taste, smell, and texture of the product they used as a child and may have come to tolerate or even like.

"Lo-Pro" Substitutes

The third component of the PKU diet—besides the protein substitutes and restricted amounts of naturally low-phe foods—consists of modified lo-pro substitutes for pasta, rice, baking mixes, and premade items such as bread, cereal, cheese, crackers, and cookies. For many people with PKU, the availability of these products has been a blessing, providing far more variety in their diet and enabling them to feel full. (Several companies that manufacture these products were established by parents dissatisfied with the choices available to their children.) The flour mixes are difficult to handle, however, and it takes time and considerable skill to bake with, for example, a lo-pro wheat or potato starch, imitation chocolate, and an egg substitute. Despite a profusion of PKU cookbooks, the failure rate of the finished product is high, especially when a child refuses to eat it.[27] Neither homemade nor commercial lo-pro products have the taste or consistency of their regular counterparts. Since the manufacturing costs for these products are high and the market small, they are also expensive (table 7.1), with shipping charges often adding to the cost. The full outlay for these products is rarely covered by health insurance, when it is reimbursable at all. Moreover, although about two hundred lo-pro products are now available, the number pales beside the approximately thirty thousand that exist for people on an unrestricted diet and allow vastly more variety in the diet.

THE CHALLENGES: CHILDREN

Difficulties in managing the PKU diet vary with a host of individual, social, and demographic factors such as the severity of disease, cost burden, degree of social support, and proximity to expert care. But perhaps the most fundamental factor is age and,

Table 7.1. Prices for Selected Lo-Pro Foods

Product	Quantity/type	Price in US dollars
Baking mix	6 lb container	29.99
Bread	6 rolls	21.99
Plain bagels	8 bagels	11.99
Butterscotch chip cookies	12 cookies	11.99
Spaghetti	17.6 oz bag	10.69
Cheese ravioli	1 lb bag	17.49
Short-grain rice	2 lb 3 oz bag	20.49

Source: Cambrooke Foods retail price list, www.cambrookefoods.com/included/docs/Retail_Price_List.pdf. The prices are as listed in 2013.
Note: The bread is produced by Artisan and the spaghetti by Aproten; other items are produced as well as sold by Cambrooke.

relatedly, whether the person has remained on the diet since infancy or came off the diet and resumed it in later life. The challenges facing children (and their parents) differ significantly in kind and degree from those confronting adolescents, while issues related to diet resumption, work, insurance coverage, and child care are largely specific to adults.

It is often difficult to persuade young children to drink the formula, and mealtimes can become a struggle. For most parents, the diagnosis of PKU came as a shock; they had no reason to anticipate a problem, and their baby seemed fine at birth.[28] Although most parents adjust well, many remain anxious about their child's health and some feel overwhelmed by their new responsibilities, especially in the early years when care demands are greatest and there is uncertainty about the child's future development.[29] The authors of a British study of feeding behavior in children aged 1 to 5 note that some parents may not comprehend the nature of the disease or the need for treatment or may lack general parenting skills. In the families studied, the children were often cajoled, bribed, or threatened to eat, and only 38 percent of the mothers reported that the children took their entire protein supplement daily. Moreover, compared with the controls, there was less meal-related verbalization in these families, such as asking if the food tasted good. The authors also note that "carers had to prepare two separate family meals, and found it difficult to serve both of them simultaneously. As a result, children with PKU were frequently fed first and given their meals alone."[30]

Following the PKU diet at school creates new and even more difficult issues. Children with PKU are expected to consume the protein substitute with lunch and cannot eat the same foods as their classmates; the diet thus marks them as different from their peers. Reflecting on their experiences at school (see table 7.2), interviewees commented particularly on the difficulties at lunch.[31] The issue was not the taste or smell of the formula, to which most were accustomed—or rather, it was not the taste or smell to *them*. Interviewees reported that if they brought lunch from home and it included lo-pro products, the sandwich bread or other modified foods seemed strange to their classmates, as did a lunch without such foods but consisting only of, say, an apple, grapes, and potato chips. Buying lunch often meant eating just the french fries and green beans, which seemed equally peculiar. Swapping foods was impossible. More generally, the inability to eat the same foods as others meant not fitting in. Nicole, age 33, commented that "for the most part, people were like, okay, well you do what you gotta do and I'm going to go have my burger, you know? And as a kid, that's hard. 'Cause you just want to have what everybody else is having." In elementary school, classmates would say about the restrictions, "Oh, that sucks. You can't have that? . . . How can you live without it?" She concluded, "If we all had PKU and we all had to do it, then it would be easy."

Some interviewees reported that their classmates and teachers readily accepted that they needed a special diet. This was atypical, however, and many more reported having been teased in school. The teasing was often so severe that the children stopped bringing the medical food to school. One woman recalled how the teasing affected her ability to follow the diet: "I quit taking the formula to school in the fourth grade because I was made fun of; it smells, and kids would hold their noses and turn their backs."

Such experiences are not uniquely American; the authors of a

Living with PKU

Table 7.2. Recollections of Experiences at School

It was a horrible thing to have the diet. Eventually I learned to hide it and keep it a secret.

You figure out you're not normal in the general scheme of things. And people want to be normal and fit in. So if you have something that makes you stand out, then you want to get rid of it.

In kindergarten, I had to have special foods and everybody was looking at me. "Why is she eating something different than us," you know? And it made me feel I didn't want to eat. I wouldn't drink my formula when I was sitting there, because I was being watched and teased and I didn't like it.

They'd say, "Your food looks and smells weird. Why does your bread look so funny?"

Kids would say, "If I see you eat one more bite, I'll gag."

Kids made fun of me at lunch; friends did, too, so that they'd fit in with the others.

British study noted that "peers can be cruel and it is not unusual for children with PKU to be taunted about their diet and forbidden foods."[32] However, there is anecdotal evidence that teasing is less common in some European countries, and since European metabolic centers are more likely than American ones to include a psychologist on their clinical team, they may be better equipped to deal with such issues.[33]

THE CHALLENGES: ADOLESCENTS AND ADULTS
Adherence

For small children whose parents manage the diet, blood phenylalanine levels are generally within the target range. But control usually deteriorates with age, with some studies indicating that as many as 80 percent of adolescents have higher than recommended blood phenylalanine concentrations.[34] Such figures must be interpreted with caution. Although there is a consensus that adherence to dietary recommendations is low in teenagers and adults with PKU, the topic has not been rigorously studied, and there is no agreement on how to define or how to measure adherence.[35] Moreover, as Anita MacDonald and colleagues note, even though patients tend to be categorized as either compliant or noncompliant (or adherent or nonadherent), compliance is actually "a dynamic process, with many patients changing between a state of compliance and partial and noncompliance."[36] That accurately describes the behavior of people interviewed for this study, most of whom were fully on-diet only as children or, in the case of women, also during pregnancy. A few participants followed all treatment recommendations and some followed none, but most compromised: they consumed the protein substitute/formula but not in the recommended quantity; or they restricted their intake of dietary protein but failed to consume the formula; or they ate the occasional slice of ordinary pizza or serving of pasta. And even those who were completely adherent at the time of the interview acknowledged having relaxed or discontinued the diet at some point in their lives.

Of course, failure to adhere to medical advice is hardly unique to PKU. Medication adherence rates in general are notoriously low, especially with chronic conditions. An oft-cited World Health Organization study concluded that, in developed countries, only 50 percent of patients with chronic diseases followed treatment recommendations.[37] And with chronic diseases generally, people tend to become less adherent over time. However, there are several specific factors associated with reduced adherence that characterize the treatment for PKU. In particular, therapy is preventive rather than curative. It involves complex dosing, with a need to consume the medical product at least three times a day and to keep track of the phenylalanine content of almost everything else that is consumed.[38] The treatment thus has a significant impact on lifestyle, an area where, in general, adherence to recommendations is typically poor and especially so when the treatment is dietary. In these respects, PKU has much in common with diabetes, where "attrition rates are high and [blood glucose] levels mostly return to baseline."[39]

Another difficulty with adherence is that people with PKU do not usually feel unwell if they go off the diet, so they lack an important incentive to comply with recommendations. Unlike people with diabetes and fluctuating glucose levels, they rarely if ever experience immediate side effects when they consume more phenylalanine than they should, so they do not get rapid feedback when they stray from the recommended diet. Even the cumulative effect may be noticeable only to friends, family, or coworkers.[40] Adults who return to the diet often remark that they only realized how much their thinking and feeling had been affected when the fog lifted and the depression, irritability, and moodiness waned.[41]

Even with relatively good adherence to treatment, there can be subtle neuropsychological impairments that interfere with planning, a skill essential to management of the PKU diet. On average, children treated early and continuously have IQ scores in the normal range, but attentional problems are common, as are hyperactivity and impulsive behavior.[42] By the fourth grade, many children with PKU begin to experience academic difficulties. A recent Spanish study found that they were twice as likely as their nonaffected peers to experience problems in school, with 39 percent requiring special tutoring and 12 percent repeating grades.[43] It is an open question to what extent these and other problems are attributable to the disease per se, to nutritional deficiencies associated with the restrictive diet,[44] to aspects of the social environment such as teasing at school, or to interactions among effects, as when neurological difficulties such as impaired attention span complicate adherence to the diet.[45] Whatever the specific cause, many adolescents and adults with PKU lack the metacognitive skills necessary to follow a complex dietary regimen.

Part of what makes PKU treatment so difficult is that maintaining stable, nonfluctuating blood phenylalanine levels throughout the day seems to be important, at least for individuals with generally well-controlled levels.[46] Ordinarily, the body regulates blood phenylalanine levels with remarkable efficiency. Irrespective of what they eat, most people maintain a stable blood phenylalanine concentration. As Cary Harding notes, "The beauty of the normal Phe control system is that we can metabolize huge amounts of Phe or very little Phe and keep roughly the same blood Phe."[47] But in PKU, where the normal metabolic process is disrupted, blood phenylalanine levels are unstable, and recent evidence suggests that fluctuations over the long term may be as important to cognitive outcomes as is the absolute level of phenylalanine exposure.[48]

Generalizations about outcomes are complicated by several factors. Many adults with PKU—a common estimate for the United States is about half—have been lost to follow-up. As they are less likely than individuals still seen at a clinic to be on the PKU diet, their exclusion from studies will inevitably bias the results. Outcome studies tend to have small sample sizes and often lack control populations, among other limitations, and their results are sometimes inconsistent.[49] There are very few data on quality of life.[50] With these caveats in mind, it can be said that outcomes in PKU are generally suboptimal. More than a third of adults studied suffer from psychiatric disorders (compared with 16.1% of controls). Emotional disturbances such as depression, generalized anxiety, phobias, and social isolation are common in PKU, as are osteopenia and osteoporosis, especially in later life.[51]

Of course, many adults with PKU thrive in their careers and in their personal and social relationships. One woman is a pediatrician, another a professor with a PhD in mathematics, a third an astrophysicist.[52] There are engineers, computer systems analysts, architects, and lawyers with PKU. Inspirational stories of adults who have excelled in school, work, and sports abound on websites and in literature by and for people with PKU. Other adults do as well as their peers without the disease. Nevertheless, a "typical profile of learning deficits and behavioural problems" has emerged from a quarter-century of research on neuropsychological outcomes in early-treated PKU.[53] Since degree of metabolic control roughly correlates with neuropsychological results, a major contributor to the disappointing statistics on outcomes is surely the difficulty of adhering to the PKU diet.

Taste

The sensation of taste for people with PKU depends mostly on whether and especially when they began eating phe-rich foods.

Early taste experiences with the low-phe diet can be surprisingly resistant to change. For example, many children resisted modifications in the medical products that were intended to make them more palatable. According to Buist, "There's something very elemental about the tastes we develop in infancy. We would find some of the kids saying, after a week or two, I don't like this [new] stuff. I want my Phenyl-Free back . . . For the most part they were happy to change, particularly I think, the younger ones. But . . . we thought that they would all just go ape for it. And they didn't all."[54]

Childhood exposure to phe-rich foods affects adult taste patterns. Interviewees who discontinued the diet as children all enjoyed some high-phe foods as adults, even if some liked steak whereas others preferred pizza, pasta, and especially real cheese (table 7.3). In contrast, adults who had continued the low-phe diet throughout childhood generally continued to avoid high-phe foods and sometimes resented changes to the formula (table 7.4).

For adults who discontinued the low-phe diet in childhood, it was difficult to later learn that they should resume the diet because clinical guidelines had changed. Once accustomed to the taste of high-protein foods, going back was not easy. There was also the emotional shock. Liz, a 40-year-old woman who was taken off the diet at age 5, commented, "And then studies came out that it would be better to stay on diet and never come off. For people who had been born in '68, who had been eating what they wanted, when they wanted, how they wanted, for however many years and then to be told at twenty-eight, hey, you know what?" Some wondered whether they had suffered irreversible damage that would perhaps manifest later in life, and all faced the question of whether to switch back to a low-phe diet.

Many people with PKU who knew about the need for a lifelong low-phe diet allowed more protein in their diet at some point, especially during adolescence or at college. The reasons had more to do with the practical and social burdens of the highly restrictive diet, however, than with a longing for the taste of chicken or beef. Indeed, these adapted taste preferences created

Table 7.3. Comments on Taste Preferences by Individuals Who Discontinued the PKU Diet in Childhood

Erin, age 33	Discontinued at age 9. Asked what she most liked to eat, responded, "Hamburgers, cheeseburgers, hot dogs, tacos, chicken."
Jessica, age 35	Discontinued at age 5, and although she could eat whatever she wanted, "It took me a long time to actually start eating things, because it took a while to acquire the taste. Like, the first meal I wanted was pizza. And we went to Papa Gino's. That's what I wanted, to go get pizza. And I hated it. I spit it back out . . . It took a while to even like a lot of the stuff . . . I still don't like ham. I still don't like milk." However, she did come to like steak, lasagne, hamburgers, pizza, and pasta, and although she tried to resume the low-phe diet at about age 32, she gave up after six months. "I didn't like the taste of the formula. And I didn't want to give up the foods I had grown to enjoy." When watching others, "I could taste it in my mouth."
Laura, age 25	Discontinued at age 5 or 6 and has cycled on and off since the age of 15. "I don't eat any meat or seafood still. Even though I can now, I don't. And I just think that it's, I never got used to a lot of those type of things. I just eat, like, fruits, vegetables, pizza . . . Now cheese is my favorite food."

a new problem: when individuals who have been treated since infancy and have never developed a taste for protein-rich foods go off the diet, they may fail to consume an adequate amount of high-quality natural protein, with the deficiency no longer offset by the amino acid supplement.[55]

"Food Is Never Just Something to Eat"

Just as sex is about more than reproduction, food is about more than nutrition—in Margaret Visser's words, "it is never just something to eat."[56] People talk and think a great deal about food; it is a pervasive aspect of our lives.[57] After all, we eat several times a day and, except for periods of ritual fasting, every day of the year. The centrality of food to our personal and social relationships and to our sense of belonging to particular ethnic and religious

Table 7.4. Comments on Taste Preferences by Individuals Who Remained on the PKU Diet after Childhood

Megan, age 34	Relaxed the diet in college. She describes taking a bite of hamburger: "I spit it out. It was just such a foreign taste and texture." She liked the traditional Phenyl-Free and was upset when its amino acid profile was changed. But the formula was and is a struggle, "as much as I like the Phenyl-Free, making sure you had three tall glasses a day . . . with school and sports and not being home all day." Conscious of the smell to others and difficulties keeping the formula cold, she wouldn't take it to school.
Kim, age 20	"Loved" her Phenyl-Free and has had a hard time since the profile was changed. She currently works at a deli, but does not find the work stressful, "because I have never had meat. So it's not like I want to eat it."
Joe, age 26	Has been consistently on the diet. When Phenyl-Free was changed, he "hated it at first. I remember I had a stockpile of the old and I wouldn't switch over."
James, age 23	Has been continuously on the diet. On people's difficulties imagining the taste preferences of those who have never eaten high-protein foods, he said, "People are shocked. They'll run through a list of everything they know that has high protein. Oh, you can't have pizza? Oh, you can't have meat? Oh, you can't have dairy products? . . . And then they think of how awful that would be, 'cause they're so used to having all those things. So they take it as a very bad thing that they'd never be able to live with. But I explain to them, if it's been like this since you were born, and you've never known anything else, it's really not so bad. And they're like, don't you ever want to have these things? Not particularly, 'cause I don't know what they taste like."

communities has the effect of making people with PKU constant outsiders at school, work, and social events. Food proscribed for people with PKU is everywhere: at the office, when friends order pizza, at birthday parties, at weddings, at baby showers, at wakes and memorial events, on celebrations of national and religious holidays—there are few social occasions, even sad ones, that do not involve food.

As a result, individuals with PKU must constantly explain their dietary restrictions. "Every time I go out to eat with new people, half the dinner conversation is about PKU," lamented one young woman. Or like this young man, they try to avoid the subject: "At a party, where all they have is pizza, I'll just tell them I'm not hungry or I already ate." People with PKU also know from experience that even if they explain the restrictions, others will sometimes forget and they will be served something they should not eat. Megan recalls returning from a friend's house and telling her mom, "I had this great snack. We should get it"—only to find out that she had been eating almonds. Friends and relatives often assume that it is okay if individuals with PKU consume just a *little* of a restricted food item, and so urge them to eat foods they know are off-limits. Thus people with PKU frequently find themselves in socially awkward situations.

Pressures from friends, colleagues, and sometimes even relatives to relax the diet may be linked to false but pervasive assumptions about the protein content of foods; after all, even many people with PKU find implausible the claim that beans and other legumes are unsafe for them to eat. The diet also challenges common assumptions about the structure of a meal. The social anthropologist Mary Douglas famously called our attention to the fact that in every culture, a collection of food items only counts as a meal if it includes certain elements. A meal cannot consist entirely of carbohydrates, or of foods that are all sweet or salty or sour or that have the same texture. To seem like a real meal, there must be multiple, contrasting elements.[58] A dinner that consisted of two vegetables and a small potato would not seem like a real meal to others.

Meals are also integral to ethnic and religious identity—as nicely illustrated by the complaint of Swaranjit Singh, a South Asian candidate for the New York City Council: "Pizza-eating people have representation. Burger-eating people have representation. Bagel-eating people have representation. But roti has no representation."[59] Eating the same foods is one way of showing that we belong to a group. More generally, meals are used to establish and cultivate relationships; sharing them is a way to create and maintain intimacy. Douglas remarks, "Even social animals in the wild use food to create and maintain their social relations."[60] Having to avoid common foods creates profound barriers to sharing meals with others, including friends, classmates, colleagues, business clients and customers, co-religionists, and members of the same Scout troop, theater club, or sports team. In part because PKU is so rare, this group identity around food is harder to establish; in most contexts, the special diet is isolating.

Adolescents, who tend to be insecure and are especially vulnerable to advertising and peer pressure, are particularly affected by the social challenges of the diet. As Michael, age 25, said in reflecting on his high school years, "It's high school. I'm just trying to fit in. The last thing you want to do is tell people you're different." Frank, age 18, described a typical situation in which everyone in a group wants to stop at a fast-food place: "I'll just get a small fry or a medium fry. Everyone else has these big burgers and I'm sitting there and I'm still hungry." Elena Vegni and colleagues, in interviews with Italian patients, found that when they were alone, they felt normal and healthy, whereas in their social relationships, they often felt uncomfortably different. Thus their lives involved a constant negotiation between feeling abnormal but being socially included and feeling healthy but being solitary. For this reason, the authors characterize PKU as a "social disease."[61]

Moreover, since every meal and snack requires planning, adolescents and adults with PKU can never be spontaneous about going to a restaurant or to a friend's house for dinner, much less on a trip. And in the United States, and increasingly in other countries, the use of prepared foods is on the rise, as are the number of meals eaten away from home. A metabolic nutritionist comments on how hard it is to adhere to the diet when eating out. She notes that many people who generally try to maintain good metabolic control eat more protein than they should because they effectively have no alternative when eating away from home. "If you're at a Chinese restaurant you gotta eat something, and you can't eat the rice . . . But that's what they have, so that's what you do. And the reality is that to restrict yourself to the degree necessary to keep your level really, really low, it doesn't work in most situations unless you bring your own food."[62] An increasing number of meals are purchased at fast-food outlets; for adolescents, these now account for a third of all meals consumed away from home.[63] In these venues, the choices are even more restricted and, from a PKU diet perspective, are usually grim, given the high phenylalanine content of the food: even a Burger King veggie burger contains 1,035 mg of phenylalanine, and one-eighth of a Domino's 14-inch cheese pizza contains 417 mg.[64]

Adolescents and adults with PKU can feel that they have little control over their lives and can resent the loss of autonomy. As Amanda, age 36, remarked, "I was very angry about having PKU. So in part, I felt like I didn't have any control over what was being said or done to me. And so I basically acted out by dumping formula down the drain and eating things I shouldn't." Such sentiments are reflected in how medical products are advertised. In her book on diabetes care, the Dutch anthropologist Annemarie Mol analyzes an advertisement for the EuroFlash, a blood sugar monitor. The ad features an image of three beautiful young people walking in the mountains. A marketing manager for the company that manufactures the device explains that the ad was intended to appeal to "people who want freedom." Mol comments, "Three young people walking in the mountains: it looks just great. The EuroFlash capitalizes on the desire of potential customers to be able to go out and walk. This walking has little to do with putting one foot in front of the other; getting a rhythm; sweating; or enjoying the wandering. What is at stake instead is the ability to walk, to go wherever you might want to go."[65] She could as easily be describing the ad for more portable forms of PhenylAde (see pp. 120–21) or the cover of a Mead Johnson brochure on PKU, which includes a composite image of four attractive adults, three actively engaged in sports: one wears a bicycle helmet, another wears riding clothes, and the third is playing tennis.[66] (The back cover includes an image of a hiker with a backpack.) As with the EuroFlash ad, what is being represented is the ability to live as one wants, not to be constantly constrained by the disease or its treatment.

Many adults with PKU, including those adhering to the low-phe diet since infancy, are also aware that the long-term effects of the disease and its treatment are unknown, and they worry about what will happen as they age. Interviewees commented on "everything that they [researchers] don't know" about the long-term consequences of dietary treatment, referring to the future as "uncharted territory." Internationally, adults often express anxiety about developing dementia, Parkinson's, or other illnesses, and they worry about how they will be cared for in a hospital or nursing home where the staff is likely to be unfamiliar with their needs.[67]

Cost

The high cost of the diet is a central concern in the United States, and it is a critical reason that many adults adhere to the diet only partially or intermittently or not at all. In the early years of screening, treatment expense was not an issue. PKU was considered an acute disease, and states generally subsidized the formula for infants and children (and continue to do so). Moreover, the FDA originally classified medical foods, such as the PKU formula, as prescription drugs, which were reimbursable for those with health insurance. In 1972, medical foods were reclassified as "special dietary foods," which were not considered drugs. This designation made it easier to develop new products for PKU and other metabolic disorders, since they no longer had to undergo costly and time-consuming clinical trials for nutritional efficacy (unless they were intended for infants under 1 year of age). But the reclassification also meant a lack of federal oversight of nutritional adequacy, and once the formula was no longer considered a drug, many insurers began to treat it as a food and stopped reimbursing for it.[68] At the same time, adolescents and even adults were increasingly advised to continue the diet, the costs of which increase with age, and to include expensive foods modified to be low in protein.

The authors of a recent review note that the "Patient Protection and Affordable Care Act requires health care plans to provide coverage without cost sharing for newborn screening services, but no national policy calls for coverage of the costs of treatment."[69] In the United States, the average cost of the medical food or formula for an infant in 2012 was about $2,275 per year, but for those in the 9 to 13 age group the annual cost rises to $8,617. Estimates for adolescent girls ($10,538) and boys ($12,483) are even higher.[70] While 38 states currently require insurers to reimburse for the medical food, the mandates often include age limitations, allow caps on the amount, or permit large co-pays. Depending on the state, mandatory coverage may exclude adults or may be restricted to pregnant women or to adult women through reproductive age. Only about half the states require insurers to reimburse any part of the cost for low-pro foods.[71] Moreover, these state laws do not apply to self-insurers, which provide at least half of all employee health insurance in the United States. According to a recent report, "denial of healthcare coverage [for medical-food treatment of PKU] is the norm for self-insured plans."[72] Of course, many people have no health insurance. Paying more than $10,000 a year out of pocket for the medical food—let alone for the expensive lo-pro products—is simply not an option for most adults with PKU.

Even when they have generally adequate health insurance, people with PKU describe disheartening struggles to obtain reimbursement. In general, insurers have little knowledge of PKU— or of any rare genetic disorder. The codes used to bill for reimbursement may require that the formula be administered by tube or be "nutritionally complete," and the codes are rarely congruent with the newer medical food products, making it easy for insurers to deny coverage. And in contrast to the situation with testing, there are no federal guidelines on treatment.[73] Even clinic visits and laboratory tests are not covered for some patients with PKU, especially adults.[74]

Metabolic clinics and health departments often make herculean efforts to assist patients in obtaining insurance reimbursement or financial assistance through state programs. (Several metabolic nutritionists interviewed for this book reported spending about a third of their time helping patients with coverage issues.) A nutritionist in a state that does not mandate insurance coverage for either the medical food or the low-pro substitutes explained that it is unrealistic to expect people to adhere to a diet they cannot afford. Recalling the heated debates in the 1980s and 1990s as to whether people should be allowed to discontinue the diet, she noted that the clinic's influence on the decision is limited, especially when the patient receives no help paying for treatment. "The patient does whatever they want to do after they get to be of age and especially when nobody has any money to help them buy the medical food. And if they haven't gotten a job and have insurance that will pay, and many insurances don't pay. So, when they get to that age and there's no money, you can't expect them to be on diet. It's just that simple."[75]

Obtaining reimbursement may be a struggle even in states with relatively expansive insurance mandates. For example, Massachusetts law requires insurers to reimburse for both the medical food and up to $5,000 per year for the low-pro substitutes. But interviewees repeatedly described a process of applying and being turned down, supplying additional documents, again being denied, and then sometimes, after several more attempts, giving up the struggle and doing without. A metabolic nutritionist commented that in Massachusetts, "where there's supposedly a law that says that formula will be paid for, it's impossible to get through the system . . . I'm dealing with a man who's in his forties who struggles but is trying to be on diet . . . He just got a bill for $10,000 . . . because the insurance company said they weren't going to pay because of some coding thing that wasn't in the contract with his particular provider. And of course, we're fighting it, but it gets really discouraging for people to have to keep fighting, and fighting, and fighting, and fighting."[76]

Financial barriers to treatment are generally less severe in Europe. A 2011 study of 16 European countries found that the cost of the medical food is entirely or largely covered in all, although low-pro foods are covered in only half, and in some cases there are age or other restrictions.[77]

The Reality of Dietary Treatment

Cost is certainly a significant barrier to treatment, and it is one that, given the political will, could be eliminated in the United States. The PKU diet is onerous even in countries where cost is no obstacle to treatment, however, because managing the disease involves many other kinds of challenges. The obstacles to strict dietary treatment are summarized by the mathematician referred to earlier, whose cognitive skills would seemingly enable her to cope with such challenges better than most:

> First, there is the amino acid formula, which must be consumed regularly throughout the day, every day, without exception. This means that you must bring (and drink!) your formula to school, work, outings, everywhere! No more impromptu meals at a restaurant or a friend's house: you must always plan ahead and prepare/bring your formula beforehand. Now, just imagine leaving for a two week vacation with your carry-on suitcase filled with several cans of powder, some of which [have] to be prepared and consumed on the plane. And this stuff smells really weird, tastes equally weird, and costs a fortune. Finally, there is the issue of having to measure all the food you eat, and convert these measurements into milligrams of phenylalanine. Then you need to add up these milligrams, and make sure you reach your target every day. While you must not go over it, you must also make sure that you get enough calories to meet your needs. In practical terms, this means that you must always carry a pocket scale, a notebook, and a small food database. The scale is to weigh the food, the notebook is to keep track of what you eat, and the da-

tabase is to obtain the conversion factors from grams of food into milligrams of phenylalanine.[78]

There is a trope in medical therapeutics—shared by many clinicians and patients—that privileges "natural" remedies over pharmaceuticals. Changes in diet seem natural and good, just as vitamins and home remedies have an emotional appeal different from taking a prescribed medication. Perhaps this partly explains why so many accounts of PKU claim that the treatment is simple. But as long as the mainstay of therapy remains a severely restrictive diet, the need for constant monitoring, recordkeeping, and planning, and especially the disruptive impact on social relationships will make it tough to adhere to treatment recommendations. The difficulties are increased and the stakes especially high for women who wish to have children. Why that is so is the subject of the next chapter.

CHAPTER EIGHT

The Perplexing Problem of Maternal PKU

In the words of a *New England Journal of Medicine* editorial in the mid-1960s, "The solution of one problem often begets new ones," and the history of PKU prevention is a case in point: "The oldest early-treated phenylketonuric girls are now more than ten years of age. Mental retardation due to phenylketonuria is waning as a result of existing programs; however, a significant incidence of retardation secondary to maternal phenylketonuria threatens unless some of these problems are solved within the next eight years."[1]

Before the advent of mass screening, most women with severe forms of PKU were childless because they were institutionalized or had been sterilized or simply because of the condition itself.[2] But the fertility of women diagnosed in infancy and placed on a protein-restricted diet is nearly normal. In the early years of screening for PKU, most clinicians assumed that it was safe to discontinue the diet in childhood, so nearly all women in the first cohort to be treated had terminated the diet by the time they became pregnant, and few continued to be seen at a metabolic clinic. Indeed, women who had discontinued the diet at a very young age might not have remembered why they had been treated or even the name of the condition.[3] Nor were systems established to track these women; after all, PKU was seen as an acute and curable condition, not, as it turned out, a chronic one. For these women, however, their fetuses were at risk because, at high levels, phenylalanine is an extremely potent teratogen. Circulating in the maternal blood of women with PKU, it is actively transported across the placenta, resulting in a maternal-fetal phenylalanine ratio of about 1:1.5.[4] Children born to such women rarely have PKU, since only about 1 in 120 offspring inherit the responsible gene from both parents. But these children were often found to be multiply impaired, although initially it was uncertain just what proportion were affected and in what specific ways and to what degree.

It is commonly said that this harm was an unanticipated consequence of the success of screening, and the magnitude of the problem of maternal PKU (MPKU) certainly did come as a shock.[5] But the possibility that damage to the offspring of mothers with PKU could be a serious complication of screening had been recognized as early as the mid-1960s, and the issue was discussed within the US Children's Bureau and in leading medical journals throughout the 1960s and 1970s. Følling himself wrote in 1971 that it is a "worrying fact that phenylketonuric mothers may give birth to mentally retarded children who suffer *in utero*, so that postnatal treatment comes too late. We may become responsible for the occurrence of more mental retardation than we prevent. At present, however, we can do nothing but continue the treatment of phenylketonuric patients, and hopefully await the further elucidation of this condition."[6] Yet discussions of dietary management only rarely addressed the issue, which was also absent from legislative hearings and the many expositions on screening in the popular press. Not until the 1980s were the risks confirmed and a prospective, large-scale study organized to examine whether damage to offspring could be prevented if women with PKU were to resume the low-phe diet.

FIRST SUSPICIONS AND RESPONSES

As early as 1956, Charles Dent, in London, called attention to the case of a woman with PKU whose three children, with unknown but presumably different fathers, were all severely mentally impaired from birth, despite having no abnormal amino acids in their urine.[7] Dent's hypothesis that the brain insult occurred in utero as the result of toxic blood phenylalanine levels in the mother made little stir at the time—not surprisingly, given that it was published in a volume of conference proceedings, antedated development of the Guthrie test, and was immediately followed by George Jervis's remark that a phenylketonuric patient under no dietary restriction at his institution had delivered two normal children. But in 1963, a group led by C. Charlton Mabry at the University of Kentucky published high-profile reports of mothers with PKU whose children were mentally retarded but did not have the disease.[8] These reports prompted Robert Guthrie to alert Children's Bureau staff and state laboratory directors to the issue and to recommend that all mothers of infants found to have high phenylalanine levels be tested for PKU.[9]

In 1966, the Mabry group reported the cases of three additional mothers (two with normal IQ) with a total of seven surviving children who were all cognitively impaired but did not have PKU. Summarizing the existing literature, the authors noted that of 50 children born to 18 mothers with the disease, 29 were mentally retarded but only 2 had PKU.[10] Many other case reports of mentally retarded but nonphenylketonuric children born to women with PKU appeared in the literature, some describing additional features such as microcephaly, cardiac and other congenital malformations, and intrauterine growth retardation, and several noting that the degree of mental impairment in the children was often more severe than that in the mothers.[11]

At the same time, a series of animal experiments established that phenylalanine easily crosses the placenta and that the phenylalanine concentration is much higher in the umbilical cord blood than in the mother's blood. Animal experiments also demonstrated that rats and monkeys born to mothers given a phe-supplemented diet suffered significant abnormalities, including learning impairments.[12] The case reports, combined with the findings from the animal studies, provoked several prominent researchers to warn of a looming problem and to urge that attention be focused on ways to prevent it.[13] Then in 1973, Columbia University epidemiologist Holger Hansen, citing Følling's recently expressed concern that the PKU program could create more MR than it prevented, published calculations for the US population indicating that the number of impaired infants born to mothers with elevated phenylalanine levels could indeed equal the numbers of those spared by screening. However, reflecting the contradictory and confusing nature of the evidence, Hansen also acknowledged that the estimates depended on unproven assumptions about the size and nature of the risk of fetal effects and that there was great variation, ranging from severe MR to normal and above-average intelligence, among the offspring of the 200 identified mothers with PKU.[14]

Investigators who were convinced that the problem was real sometimes proposed reducing the amount of phenylalanine in the diet of women with PKU during pregnancy. It seemed at least plausible that, since brain damage in PKU could be prevented through control of blood phenylalanine level, such control in pregnancy might also prevent teratogenic effects in the fetus. As early as 1964, an editorial in the *Lancet* advised that "we ought to begin to think now about pregnancy in treated phenylketonurics."[15] Soon after, an editorial in the *New England Journal of Medicine* warned that the success of screening could be undermined by an increase in MR secondary to maternal PKU and urged development of a low-phe diet that would be palatable to adults (noting, with considerable understatement, that the existing formula was unlikely "to be accepted with pleasure by patients who have grown accustomed to normal food").[16] Beginning in the mid-1960s and continuing through the 1970s, clinicians were often urged to restrict the phenylalanine intake of pregnant women with PKU.[17] But at the time, there were substantial obstacles to systematically acting on this advice.

The first was the difficulty of identifying women with PKU who were approaching childbearing age, most of whom had not been seen at clinics for many years. One clinician noted that "with the diet being stopped at the age of five or six and the far more fragmentary services back in the '60s and '70s than there are today, these patients were dropping through the cracks like sand out of a sieve . . . We had treated them. We had cured them. And so follow-up of these people was very, very, sparse."[18] And the women themselves were unlikely to understand the gravity of the situation, so would not have communicated the risk to their obstetricians.

Then, even if the women could be identified, the onerous nature of the diet, especially during pregnancy, made adherence unlikely. Lofenalac was designed for use by infants, not older children or adults. As a nutritionist who worked with Willard Centerwall in the 1950s explained, "an infant who starts as a newborn on the diet, they'll sit down and eat out of the can with a spoon as they get older because they are accustomed to that taste as well as the odor. But the older children, if they hadn't started taking it within the first three months of life, you practically had to starve them into taking the product because it tasted so different."[19] Most adults who had become accustomed to eating normal foods were likewise unable to tolerate the taste of the formula.[20] Indeed, in the 1960s, there was but a single reported successful attempt at resumption of the PKU diet during pregnancy.[21] Every other reported effort at resuming the diet ended in failure.[22]

Because it takes time to reduce blood phenylalanine levels, some researchers hypothesized that to prevent fetal harm, the diet should be resumed early, even before conception. Thus, pregnancies would have to be carefully planned or the diet reintroduced when a woman reached her childbearing years or at marriage.[23] Such recommendations seemed unrealistic, given the difficulties in following the diet. In addition, women who had discontinued the diet as children generally experienced some degree of cognitive and neuropsychological impairment, which further eroded their ability to plan and implement a difficult task such as preconception dietary management.

Complicating the picture still further, it was impossible to know, given the dearth of successful cases, whether the diet would actually prevent MR. Hansen even questioned whether high maternal blood phenylalanine was a cause of cognitive impairment in offspring. The first researcher to control for maternal IQ, he found a strong correlation between the child's and the mother's intelligence, but not between the child's intelligence and the mother's phenylalanine level. Hansen thus hypothesized that, though dietary treatment might forestall cardiac and other fetal harms, it would not prevent MR in the offspring of mothers who were themselves cognitively impaired.[24]

Some influential researchers thought the PKU diet could result in nutritional deficiencies that were harmful to the fetus and possibly also to the mother. Animal researcher Henry Waisman, whose work had demonstrated the damaging effect of maternal hyperphenylalaninemia in rat and monkey offspring, suggested that pregnant women with PKU should restrict phenylalanine intake only if they had previously given birth to a severely retarded child.[25] Biochemist Samuel Bessman, whose animal work demonstrated that restriction of phenylalanine in pregnant rats resulted in liver abnormalities in the litters, likewise believed that treating pregnant women with a low-phe diet increased the risk of brain damage in their children.[26] Nor was it clear what proportion of infants of affected mothers were actually at risk, or why the offspring of some mothers were severely affected while others were apparently unharmed. Early studies indicated that only about a third of the offspring of mothers with PKU were, to some extent, cognitively impaired, and there was apparently no increase in the incidence of nonphenylketonuric MR among relatives of individuals with PKU.[27] The relationship between maternal phenylalanine levels and the frequency and type of fetal effects in the offspring was also murky.[28] Moreover, ascertainment bias in these studies was a major concern, since cases typically came to notice because of abnormalities in the child. In short, the situation in the 1960s and 1970s was characterized by extreme uncertainty as to the kind and frequency of fetal harm, its cause, its relationship to the severity of the mother's PKU, its timing in fetal development, and whether it was preventable.

As early as 1977, some PKU experts, including Richard Koch, responded to concerns about MPKU and to reports that IQ scores of individuals with PKU fell when the low-phe diet was discontinued by recommending that the duration of treatment be extended for children and young adults.[29] Given the uncertainty about MPKU, however, it is not surprising to find a lack of consensus on what, if any, action to take. Some public health leaders were also influenced by earlier criticism of mandatory newborn screening for PKU. The US Children's Bureau, in particular, had been accused of prematurely promoting universal screening, and its harshest and most persistent critic was Bessman, who now urged caution with regard to the treatment of MPKU. Bureau staff, once burned, were understandably reluctant to promote an unvalidated treatment to solve a problem whose seriousness and scope were as yet unknown. Their concerns are strikingly expressed in an internal memo from Mary Egan, chief of the Nutrition Section, in response to the articles by the Mabry group and the editorial in the *New England Journal of Medicine*. Egan wrote, "Personally, I feel we ought not to circulate material which advocates a means of treatment when such means has not been adequately studied." She faulted the journal's editorial for advising treatment of pregnant women "when we have no information about how they ought to be treated," and she went on to say that "I have become a great conservative in this area since I have been involved in 'defending' [the] Children's Bureau stand in PKU and other metabolic diseases. The implications of 'widespread distribution' of prematurely drawn conclusions have been criticized—and rightfully in my estimation. We have a responsibility not only to be 'purveyors of information relating to child health and welfare' but also to use judgment in what we 'purvey' and to foresee the results of such activity."[30]

FAMILY PLANNING AS A RESPONSE TO MPKU

Given both the challenges of maintaining metabolic control during pregnancy and its very uncertain benefits, clinicians often urged women with PKU not to have biological children.

This meant avoiding pregnancy and, following the US Supreme Court's 1973 decision in *Roe v. Wade*, having an abortion in the case of an unplanned pregnancy. Thomas Perry, a leading metabolic researcher at the University of British Columbia, Vancouver, was an especially strong advocate for the position that women with high blood phenylalanine levels should not become pregnant and, if they did so, should undergo therapeutic abortion.[31] In 1973, Hansen reported that "avoidance or termination of pregnancy in all women with elevated blood phenylalanine appears to be the usual recommendation."[32] Indeed, an influential set of guidelines for screening published by the World Health Organization asserted that "until the efficacy of a low protein diet during pregnancy is proven, termination of pregnancy should remain an option."[33]

Not everyone agreed that women with PKU needed to avoid having children. Some clinicians were more optimistic about both the efficacy of a phe-restricted diet and women's ability to adhere to it, or they believed it was not their decision to make.[34] A psychologist long involved with MPKU noted that few women ever asked whether they should get pregnant. A small number came to the clinic when planning pregnancies, and it was assumed that they would do well. But most women were seen at the clinic when they were already pregnant, and they wanted their babies. "We as a clinic did not do what many clinics did, which was to say terminate and try again . . . we weren't comfortable with that." Instead, clinic staff tried to instill the message that young women should not get pregnant unless they were willing to plan their pregnancies. But such planning requires forethought, a capacity often diminished in women with high blood phenylalanine levels. When the advice failed, and women made their first visit to the clinic when already pregnant, clinicians tried "to support them in whatever they decided."[35]

Supporting women in their choice to have children became more common with increasing confidence in the efficacy of a low-phe diet in MPKU, improvements in the medical foods, changes in guidelines regarding diet discontinuation, and a general cultural shift in attitudes about reproductive decision making. Reflecting this shift to greater patient autonomy, a nutritionist who, for decades, has counseled many women with PKU described a 2007 clinical encounter involving the unplanned pregnancy of a woman whose previous (uncontrolled) pregnancies resulted in the births of two cognitively impaired children, one of whom had been removed from her custody. The woman said she planned to terminate this pregnancy, and the clinician thought to herself, "In her case, having another child that she can't manage is not a good idea. But that's my mind, not her mind. So I told her, it sounds like you really gave it a lot of thought and you're trying to come to a mature decision. And whatever you decide, we'll support you . . . If you change your mind and decide to keep it, let us know, we'll help you through the pregnancy. But I can't be telling people how to run their lives."[36] In the first decades of screening for PKU, however, most clinicians advised women with PKU not to have children, given the many uncertainties surrounding the nature and extent of the problem, the efficacy and practicality of treatment, and the possibility that the fetus might suffer nutritionally if the mother's diet were too restricted. As late as 1980, the authors of one study flatly concluded, "At present, females with classical and variant forms of PKU are advised not to become pregnant for fear of fetal damage."[37]

THE MATERNAL PKU COLLABORATIVE STUDY

The North American Maternal PKU Collaborative Study (MPKUCS) was launched in 1984 at several centers in the United States and Canada, and later in Germany. Given that the first alarms about MPKU had been raised from the very start of screening, why did it take two decades to mount a research effort to discover whether the problem was real and, if so, whether it could be alleviated by treatment? The main reason is that the very small community of PKU researchers and clinicians was focused on other immediate questions, particularly whether treatment of individuals with PKU was always necessary. As discussed in chapter 5, during the 1960s and 1970s, no consensus existed as to what proportion of individuals who received a positive result on the Guthrie test were at risk of cognitive impairment, and thus there was considerable concern that many were being treated unnecessarily. There were other pressing questions, too. How well did the diet work? How long did it need to be maintained? Was it adequate for normal growth and development? Compared with such pressing issues, maternal PKU seemed a distant problem. Given that statewide screening programs only started in 1963, for example, few treated women had yet reached reproductive age. Those who were ready to have children were usually no longer on the protein-restricted diet and had been lost to follow-up and thus could not easily be studied. In the 1960s and 1970s, instances of MPKU were so rare that each one justified publication of a case report. Despite the concerns raised by some experts, no one knew for sure whether the problem was large or small.

That situation changed dramatically in 1980, when Roger Lenke and Harvey Levy published the results of a large survey of treated and untreated pregnancies of women with elevated phenylalanine levels. They found that 92 percent of the offspring of untreated women with classical PKU met the criteria for MR. Moreover, the frequency of microcephaly in the children was 73 percent, with smaller but significant risks of congenital heart disease and low birth weight.[38] Levy and others recognized at the time that ascertainment bias and the inclusion of cognitively impaired women born before newborn screening for PKU might have slightly inflated the figures, and it soon became evident that the risks to the fetus were much lower for mothers with mild and atypical phenylketonuria.[39] But with respect to classical PKU, the rate of mental retardation remained very high even when the MPKU was ascertained from screening, and there was no statistically significant correlation between the IQs of mothers and children. Moreover, the mean IQ of the children of relatively high-IQ mothers with PKU was only 57. Thus, it seemed that the risk of MR resulting from fetal exposure to high phenylalanine concentrations was at least as high as the risk in PKU.[40]

Both the frequency and the extent of damage to offspring reported by Lenke and Levy astonished the community of metabolic researchers. Two years later, Henry N. Kirkman used Lenke and Levy's data to project a "rebound" effect in PKU, which indicated that, in the absence of effective measures to address the problem, the frequency of MR in infants of mothers with PKU might completely offset the reduction in frequency of PKU-related MR gained by newborn screening and treatment.[41]

In his 1982 article, Kirkman considered two mechanisms that could potentially produce a rebound in the frequency of MR: (1) an increase in the incidence of the responsible gene and (2) the phenomenon of MPKU. In a reprise of the argument made by Penrose in his 1946 inaugural address, Kirkman claimed that the potential impact of the first was negligible, since policies aimed at reducing the frequency of the offending gene can only target affected homozygotes. As with any rare, autosomal recessive condition, the vast majority of the offending genes are carried by symptomless heterozygotes. Hence a program of eugenic selection could only very slowly reduce that gene's frequency. And because mutation is rare, the relaxation of selection—in the case of PKU, through screening and treatment—would only very slowly increase the frequency. Thus the dysgenic effect of dietary therapy for PKU would be inconsequential, at least in the next few generations.[42]

As to the second mechanism, the potential impact of MPKU on the incidence of mental retardation would be both sudden and severe. Kirkman suggested visualizing the problem as follows. Consider, in the days before treatment, a man and a woman with PKU; neither of them has children, because they are severely impaired and institutionalized. In the next generation, consider a man and a woman who have PKU but, because of screening and treatment, are not impaired. For the man, the chance of his fathering a child with PKU is small, but even if he does so, the child will be treated. The woman, if not on a protein-restricted diet, can be expected (because of MPKU) to have, on average, two impaired children. Thus, in both the first and second generations, there are two cases of PKU-related MR.[43] Kirkman noted that the height of the rebound would be affected by several factors, including the average gravidity of women with PKU, the rates of fetal and postnatal survival, and the percentage of pregnancies in which at least a degree of dietary control is maintained. He also warned that if the problem were not addressed, the results could adversely affect public perceptions of programs to study and prevent MR and genetic diseases, and he noted that no coordinated plan existed to deal with the problem.[44] The combination of Lenke and Levy's survey and Kirkman's epidemiological projections, now based on concrete statistics regarding the magnitude of fetal damage, galvanized the community of PKU researchers.

A definitive MPKU study was nearly derailed even before it began, however. As plans were being considered at the National Institute of Child Health and Human Development to launch a large prospective trial, geneticist Savio Woo and his colleagues isolated and cloned the human phenylalanine hydroxylase (PAH) gene.[45] Their 1983 publication suggested to the NICHD director that knowledge of the PAH gene would quickly lead to definitive therapy through genetic manipulation, replacing the dietary approach to PKU and making questions about the safety and efficacy of the protein-restricted diet in pregnancy obsolete—and thus rendering the MPKU study unnecessary.[46] Although many shared his optimism about gene therapy, the director did not prevail, and in 1984 the prospective, longitudinal MPKUCS was initiated to establish the efficacy of a restricted diet in preventing maternal PKU. The study was also designed to address several subsidiary questions. What was the optimal level of phenylalanine during pregnancy? Did the diet provide for adequate fetal growth and development? To be effective, should the phe-restricted diet be started prior to conception? Was it necessary to treat pregnant women who have mild hyperphenylalaninemia?[47]

The initial results from the MPKUCS were disconcerting. Only about a quarter of the 402 women studied were on a phe-restricted diet when they became pregnant; 10 percent achieved the recommended level of metabolic control by 10 weeks' gestation, and a mere 5 percent achieved control prior to conception. Most women in this first cohort of individuals identified by newborn screening would have discontinued the diet in childhood, making resumption difficult and affecting their cognitive and neuropsychological function. In the study group, for example, 25 percent of the women had IQs below 80; in contrast, in a random group of women, fewer than 10 percent would score so low on a cognitive test. It is likely that an even larger percentage of women in this first cohort would have had deficits in executive function that undermined their ability both to plan pregnancies and to maintain metabolic control. As a result, when the study began, there were few outcomes to report for women treated before conception or with well-controlled pregnancies.[48]

Women in later cohorts of the MPKUCS had remained on the diet longer as children, at least to some degree, and their abilities to plan for and cope with pregnancies increased. The struggle to achieve metabolic control before and during pregnancy was also eased by a wider array of better-tasting, more convenient, and easier-to-prepare low-pro products. Results of the MPKUCS and later studies eventually confirmed the efficacy of the diet in reducing the risk of cognitive impairment, cardiac defects, low birth weight, and microcephaly in the offspring of these women. The results also supported the association of earlier metabolic control with better outcomes in offspring, the value of keeping blood phenylalanine concentrations low throughout pregnancy, and the value of resuming the diet before conception. Achieving phenylalanine control prior to pregnancy is critical because many organ systems are already developing before a woman knows she is pregnant; pre-conception control reduces the risk of cardiac defects, in particular.

ONGOING CHALLENGES FOR WOMEN WITH PKU

Despite the gains since the 1980s, women frequently report that maintaining a low-phe diet during pregnancy remains a formidable task. Since phenylalanine levels that are safe for the mother may damage the fetus, pregnant women are advised to consume even less protein than they otherwise would and to drink about

25 percent more formula. Tolerance for the formula is typically lower during pregnancy, especially in the first trimester, when nausea and vomiting are common.[49] Erin, a woman in her early thirties with two children, noted that she would try to drink three glasses of formula a day, but "half the time, I almost threw it up. Sometimes I cheated. I drank half and I dumped the rest to make people think I was drinking it."[50] Even today, many women of reproductive age discontinued the diet as children, and most others relaxed it during adolescence. The need to consume less protein thus means giving up foods to which they have become accustomed. Lisa, a woman in her early forties who had continued drinking the formula after childhood but was otherwise eating normally, described the struggle: "Unfortunately, there's just some foods I couldn't give up. I was going crazy not having them—the steak, the hamburgers, the pizza. I love spaghetti and the sausage and the meatballs and stuff like that so it was tough, 'cause I'm watching everyone else make it and I could taste it in my mouth. And it's just like I have to have it."[51]

The usual challenges of staying on the PKU diet are exacerbated by the need to gain weight during pregnancy. Some women point out that it is difficult to eat more when they are filled up by the formula; others find that the foods they are allowed to eat have no appeal. The foods a doctor recommends for morning sickness often contain too much phenylalanine. Friends and acquaintances sometimes comment that the baby must not be getting enough nutrition. Other women describe often being hungry when pregnant but allowed no more phenylalanine for the day. Table 8.1 gives a sense of the daily diet of a woman adhering to the guidelines for food intake during pregnancy.

Apart from challenges with the PKU diet, pregnant women also face the added pressure of constant waiting for and worrying about blood test results, instead of just enjoying the pregnancy as they see many of their friends doing. Although health insurance companies are more likely to pay for formula and medical foods during pregnancy, actually receiving payments is often no mean feat. And of course, pregnancy does not relieve women with PKU

Table 8.1. Sample Daily Intake Recommended for Pregnant Woman with Maternal PKU

12 oz of formula
Breakfast: Low-protein toast with jelly and margarine; grapes
Snack: Low-protein cookies
Lunch: Sandwich with lettuce, tomato, mayo, pickles on low-protein bread; apple
12 oz of formula
Snack: Apple
Dinner (out): Pizza with no cheese—1 slice; side salad with Italian dressing; Italian ice
Before bed: 12 oz of formula; popsicle

Source: Sample diet provided by Frances J. Rohr, metabolic nutritionist, Children's Hospital Boston.

from the challenges of daily life. Women who are responsible for preparing meals for the family typically cook food they themselves cannot eat. Many women do most or all of the housework, are responsible for child care, and hold a job (essential for insurance coverage). Finding the energy to plan and cook special meals, prepare the formula, and keep track of the diet can be especially difficult at certain stages of pregnancy.

With these daunting challenges, it is not surprising that, in the United States, only about a third of women with PKU (of the perhaps half who attend clinics for follow-up) are on the protein-restricted diet at the time they become pregnant. By 10 weeks' gestation, about 55 percent of these women do achieve metabolic control.[52] That is certainly an improvement over the situation in 1994, when the first results of the MPKUCS were published. The height of the rebound in MR was much less than Kirkman had projected as likely to occur in the absence of efforts to address the problem of maternal PKU.[53] This outcome is attributable to the use of therapeutic abortion in untreated pregnancies, the generally lower severity of cognitive impairment in MPKU than in untreated PKU, and efforts to help women achieve metabolic control before or early in their pregnancy.

Most women who do manage to resume the PKU diet during pregnancy discontinue it immediately after the birth of the baby.⁵⁴ The experience of Kristen, a woman in her late thirties with two children, is typical. She explained that on deciding to have children, she resumed the diet and maintained it until her first child was born, "then I came off of it. And then when I decided I wanted another child, I went back on the diet, got pregnant, went through it, and then I came back off the diet. So I've pretty much just been on the diet when I'm pregnant."⁵⁵ Discontinuation of the diet by women with PKU is associated with increased thought and mood disorders and other psychosocial problems that may impair a mother's ability to care for her child and may adversely affect the home environment. As in studies of other prenatal exposures to toxins, there is evidence that the quality of the home environment has a significant effect on developmental outcomes in children of mothers with PKU, perhaps as great as the timing of maternal metabolic control in pregnancy.⁵⁶

Currently, there is much interest in nondietary treatment of PKU, especially for pregnant women. Clinical trials using gene therapy, hepatocyte transplantation, or enzyme replacement drugs are increasing, and pregnancy is likely to be one indication for health insurers to pay for expensive treatments such as sapropterin. Enthusiasm for such medications may dampen interest in studying the mundane obstacles to effectively managing the diet, especially during pregnancy. Indeed, despite the continuing problems of maternal PKU, the most recent study of developmental outcomes of offspring in cases of MPKU appeared in 2003. This small Australian study found that most women did not achieve metabolic control until after the first trimester and that poor control correlated with poor cognitive and behavioral outcomes in their offspring. Reprising the warnings of Følling, Hansen, and Kirkman, the authors concluded that in the absence of a system to routinely manage pregnancies and provide follow-up for the children, "we are in danger of negating all the gains made over the last three decades from neonatal screening for PKU."⁵⁷

CHAPTER NINE

Who Should Procreate?
Perspectives on Reproductive Choice and Responsibility in Postwar America

When Savio Woo and his colleagues announced, in 1983, their discovery of the gene whose mutations result in PKU, they titled their paper "Cloned Human Phenylalanine Hydroxylase Gene Allows Prenatal Diagnosis and Carrier Detection of Classical Phenylketonuria."[1] Rather than highlighting the potential for gene therapy, the authors concluded by predicting that because of their work, "prenatal diagnosis of classical PKU and carrier detection should become a reality for most of the PKU families in the general population." The focus on the reproductive implications of their research may seem surprising, especially given the frequent deployment of PKU in critiques of eugenics. Furthermore, many of the parent-professionals who passionately advocated for newborn screening for PKU, such as Robert MacCready, Robert Cooke, and Elizabeth Boggs, rejected sterilization as a means of prevention, as did the President's Panel on Mental Retardation and the NARC.[2] Robert Guthrie's life was dedicated to preventing MR through environmental interventions—indeed, his other major campaign was eradication of childhood lead poisoning—not through avoidance of the birth of affected individuals.[3] But while newborn screening advocates abjured eugenics, the reproductive issues that loomed so large for PKU researchers in the pre–World

War II period would continue to be salient for some scientists, and recognition of the problem of maternal PKU would prompt many clinicians to advise women with PKU to avoid having biological children.

The concern with carrier detection and the strongly directive reproductive counseling may seem at odds with the popular view that eugenics had long been discredited. On the conventional view, eugenics fell victim first to advances in genetics, which subverted its scientific assumptions, then to the Great Depression, which undermined the equation of social status with genetic worth, and finally and fatally to revelations of Nazi atrocities, which destroyed its moral authority. As Alison Bashford has recently noted, in this conventional popular and often scholarly narrative, the Nuremberg doctors' trials of 1946–47 and the United Nations' convention on genocide and Universal Declaration of Human Rights of 1948 (signed by 30 nations by 1967) "collectively marked the end of eugenics."[4] With its demise, the discourse of reproductive autonomy ostensibly replaced talk of the need to restrict reproductive choices on behalf of children-to-be or the larger society. In the new perspective, procreation was a private affair, there were no right or wrong reproductive decisions, and everyone had the right to start a family or otherwise fulfill their reproductive desires, whatever they might be. Or at least, this is a standard account of the fate of eugenics in the postwar era.

Of course, not everyone agrees that eugenics died—either in the 1940s or later. In particular, there is an important line of argument according to which contemporary medical genetics constitutes a new type of what is variously called "backdoor," "private," or "consumer-oriented" eugenics.[5] In this perspective, the agents of eugenics are potential parents. Responding to social norms of health, intelligence, attractiveness, and so on, they *want* to use genetic technologies to control the characteristics of their offspring. Thus, the view that contemporary medical genetics represents a reincarnation of eugenics does not challenge the usual chronology, according to which a commitment to procreative liberty replaced the belief that reproductive decisions should be subject to social control. Indeed, if anything, the assumption that contemporary reproductive genetics is "all about choice" reinforces that chronology.

Yet this standard periodization coheres neither with the concern about PKU carriers that preoccupied researchers following discovery of the disease and reemerged in the 1960s nor with the strongly directive reproductive counseling of young women that characterized the 1960s, 1970s, and even early 1980s. Clearly, premises and practices generally thought to have been discredited by the mid-1940s persisted in at least some professional domains far longer than is usually assumed. Ever since Lionel Penrose's famous inaugural address at University College London, the case of PKU has served to exemplify what is wrong with eugenics. But the ability to detect and treat the disease actually *intensified* concerns in some quarters about the human genetic future. The case of PKU helps illuminate the wider context of changing norms about the regulation of reproduction in postwar America.

IN THE AFTERMATH OF WORLD WAR II

As Dorothy Nelkin and Susan Lindee note, in the immediate aftermath of World War II, revelations of Nazi atrocities produced widespread revulsion against genetic explanations of individual and group differences, with the pendulum swinging from a hereditarian to an environmentalist perspective. They also observe that this shift from biological to cultural determinism was short-lived and that, even in the 1950s and 1960s, eugenic language persisted among professionals in fertility and in human genetics.[6]

Indeed, something of a backlash among geneticists was evident shortly after the war's end. Classical geneticists such as H. J. Muller and Julian Huxley, who had campaigned for programs of selective breeding before the war, returned to the fray. In 1947, Muller was awarded a Nobel Prize for discovery of the mutagenic properties of X rays, a development that allowed him to speak with enhanced authority on genetics-related social issues. Two years later, in his presidential address to the newly founded

American Society of Human Genetics, Muller argued that the human species was deteriorating under an ever-increasing load of deleterious mutations. In his view, this burden was attributable both to expanded military and medical uses of radiation (especially atmospheric nuclear testing) and to therapeutic advances in medicine, which allowed individuals who once would have died before their childbearing years to survive and reproduce. To counter this degeneration, he proposed a scheme under which the most burdened 3 percent of the population would voluntarily refrain from reproducing.[7] Several years later, he also resurrected a proposal of the 1930s to bank the sperm of men outstanding in intellect, temperament, and character, a program then renamed "germinal choice" to emphasize its voluntary character.

Muller's warnings about genetic deterioration and hopes for genetic improvement appear to have resonated widely; numerous echoes of his argument are found in both popular and scientific literature. His impact is reflected in a colleague's comment that Muller's contribution to a symposium titled "The Control of Human Heredity and Evolution" in 1963 "was received with such enthusiasm by the audience that, after hurried consultation with some of the other speakers, we agreed it would be anticlimactic and quite undesirable to throw it open for general discussion."[8] Interest in the control of human evolution was also spurred by discovery of the double-helical structure of DNA and unraveling of the genetic "code." Molecular scientists such as Francis Crick, Joshua Lederberg, Salvador Luria, Linus Pauling, and Robert Sinsheimer began to debate the pros and cons of what Rollin Hotchkiss in 1965 termed "genetic engineering."[9]

While some molecular scientists, such as Luria, urged caution, many agreed with Muller's pessimistic view of the future. Lederberg, who was generally skeptical of schemes to control human breeding, asserted at a 1962 symposium "Man and His Future" that "the facts of human reproduction are all gloomy—the stratification of fecundity by economic status, the new environmental insults to our genes, the sheltering by humanitarian medicine of once-lethal defects."[10] Francis Crick went further. At the same symposium, Crick expressed his agreement "with practically everything Muller said" about the urgent need both to prevent further genetic deterioration and to increase the proportion of superior genotypes in the population. In place of the current laissez-faire system of reproduction, Crick argued, we might substitute a licensing scheme, whereby "if the parents were genetically unfavourable, they might be allowed to have only one child, or possibly two under special circumstances." Commenting on Crick's suggestion, British biochemist and virologist N. W. Pirie asserted that, on the question of whether there is a right to have children, "in a society in which the community is responsible for people's welfare—health, hospitals, unemployment insurance, etc.—the answer is 'No.'"[11]

Of course, Muller's scientific assumptions and policy proposals were also strenuously contested, with some prominent geneticists emphasizing the slow rate of genetic deterioration. Thus, Charles Scriver, perhaps the world's leading authority on inherited metabolic disorders, asked, should we "worry about the accumulation of mutant genotypes in the future as the ultimate legacy of 'good treatment' today?" and answered, "The rate of genetic deterioration accountable to such accumulation would be small and hopefully the rate of accumulation of new knowledge to solve the dilemma will exceed the former rate."[12]

Whether Muller's fellow scientists endorsed or rejected his claim of an urgent need to address the problem of genetic deterioration, most would have agreed that those who knowingly transmitted a serious genetic disease were, at the very least, morally culpable. For example, in a book that took issue with virtually every other aspect of Muller's eugenics, the geneticist Theodosius Dobzhansky remarked that persons who carry serious genetic defects should be persuaded not to reproduce and that, if persuasion should fail, "their segregation or sterilization is justified." He continued, "We need not accept a *Brave New World* to introduce this much of eugenics."[13] The anthropologist Ashley Montagu, who also criticized eugenics and was unconcerned with the quality of the gene pool, wrote that "there can be no question that infantile amaurotic family idiocy [Tay-Sachs disease] is a disorder that no one has a right to visit upon a small infant. Persons carrying this gene, if they marry, should never have children, and should, if they desire children, adopt them."[14] Likewise, according to Sheldon Reed (who coined the expression "genetic counseling"), "The most important counseling problem [in PKU] will be trying to educate affected individuals so that they will not wish to marry another affected person or a carrier . . . No couple has the right to produce a child with a 100 per cent chance of having P.K.U., and it is doubtful whether a couple has the right deliberately to take a 50 per cent chance of producing such a serious defect."[15] What mattered for these scientists was not the distant future but the potential short-term consequences, both for the child-to-be and for the larger society. And they agreed that these consequences were appropriately a matter of social concern.

In theology as well, the emphasis on disputes regarding specific eugenic solutions can obscure the fundamental agreement on the view that some people should not reproduce. In the 1960s, Protestant thinking on bioethical matters was dominated by two theologians, the Episcopalian Joseph Fletcher and the Methodist Paul Ramsey.[16] The views of Fletcher and Ramsey are often counterpoised, for they seem to have disagreed about everything. Fletcher was a Civil Rights and anti–Vietnam War activist who marched and picketed for many liberal causes, whereas Ramsey was a social conservative who defended the war. They also sparred over the ethics of abortion, euthanasia, sperm banking, artificial insemination by donor, cloning, and other real or potential genetic manipulations. Fletcher, the founder of "situation ethics," was a utilitarian who judged the rightness or wrongness of acts by their consequences for human well-being. From this standpoint, humans should aggressively take charge of their own evolution. "Man is a maker and selecter and a designer," he wrote, "and the more rationally contrived and deliberate anything is, the more human it is." This perspective was reflected in the title of his 1974 book *The Ethics of Genetic Control: Ending Reproductive Roulette*. Ramsey strenuously disagreed. In his deontological perspective, life was sacred, and abortion, euthanasia, and all potential forms of human genetic engineering were inherently wrong.[17]

Despite their very real disagreements, both strongly endorsed negative eugenics. Thus, Fletcher maintained that individuals at risk for transmitting grave diseases should not be allowed to reproduce. "The right to conceive and bear children has to stop short of knowingly making crippled children . . . just as the rights of parents have had to bow to required schooling and the rights of voluntary associations have had to bow, in public services, to the human need to be respected regardless of ethnic and racial differences."[18] Fletcher sometimes put the point even more strongly. As late as 1980, he claimed that "reproductive rights are not absolute and those who are at risk for passing on clearly identifiable, severely deleterious genes and debilitating genetic disease should not be allowed to exercise their reproductive prerogative." After all, "testes and ovaries are communal by nature, and ethically regarded they should be rationally controlled in the social interest."[19]

On this point at least, Ramsey agreed. In *Fabricated Man*, his 1970 critique of genetic engineering, Ramsey maintained that Christian teachings have always held procreation to be a duty to future generations, not a matter for the "selfish gratification of would-be parents." No one has a right to have children for his or her own sake, he argued. Indeed, the marriage licensing power of the state might reasonably be used to prevent the transmission of grave diseases. To those who would consider this policy an infringement of liberty, he replied that "it ought never to be believed that everyone has an unqualified right to have children, or that children are simply for one's own fruition . . . the freedom of parenthood is a freedom to [exercise] good parentage, and not a license to produce seriously defective individuals to bear their own burdens."[20]

THE "POPULATION BOMB"

Another important if indirect factor in the revival of eugenics was anxiety over a perceived population explosion resulting from agricultural improvements, public health measures, and the development of antibiotics such as penicillin and streptomycin. At numerous conferences and in a raft of scholarly and popular books and articles, the argument was that if world population were not kept in check, it would be impossible to maintain a minimum standard of living. According to the massive report *Biology and the Future of Man*, the work of the Committee on Science and Public Policy of the National Academy of Sciences, "the upsurge in the growth of human populations constitutes the major problem for the immediate future of man."[21] In his 1968 blockbuster book *The Population Bomb*, ecologist Paul Ehrlich predicted imminent famine on a massive scale and suggested adopting various financial rewards and penalties to discourage reproduction; these included increasing the amount of family income that was taxable with each child, imposing luxury taxes on cribs and diaper services, and awarding "responsibility prizes" to couples for every five years that they remained childless.[22] The sense of alarm surrounding the issue is reflected in economist Kenneth Boulding's proposed system of marketable licenses to have children: "Each girl on approaching maturity would be presented with a certificate which will entitle its owner to have, say, 2.2 children or whatever number would ensure a reproductive rate of one. The unit of these certificates might be the 'deci-child,' and accumulation of ten of these units by purchase, inheritance, or gift would permit a woman in maturity to have one legal child. We would then set up a market in these units in which the rich and the philoprogenitive would purchase them from the poor, the nuns, the maiden aunts, and so on."[23] (He thought the plan had the added advantage of reducing income inequality, since the rich would have more children, leaving them poorer, while the poor would have fewer children, leaving them richer.)

Boulding aimed to solve the population problem while preserving a maximum of individual choice. But coercive proposals were also common, justified both by the gravity of the situation and by the assumption that limiting population growth is a public good unachievable through an appeal to individual families with economic or emotional interests in having more children. The concept of a "tragedy of the commons"—situations in which individuals acting rationally produce an outcome in which everyone is worse off—was popularized by ecologist Garrett Hardin, who argued that "the only way we can preserve and nurture other and more precious freedoms is by relinquishing the freedom to breed, and that very soon."[24] It was but a short step to the conclusion that if breeding must be limited, the restrictions should be selective, particularly since the population control movement had deep roots in the pre–World War II eugenics establishment. Indeed, all the main organizations and individuals promoting population control in the 1950s and 1960s, such as Frederick Osborn, Kingsley Davis, Frank Notestein, Robert C. Cook, and Clarence Gamble, had earlier supported eugenics research and advocacy.[25] It was only natural for them to ask why procreative liberty should not be limited for the purpose of improving the population if it could legitimately be limited for the purpose of reducing its growth. As Vance Packard summarized a common argument of the day, "If you are going to try to control the quantity of the population, why not also control the quality?"[26]

MEDICAL PROGRESS AND GENETIC DETERIORATION

Worries about the long-term genetic impact of medical progress were certainly not new. After noting in *The Descent of Man* (1879) that doctors do their best to save every life, Charles Darwin commented, "There is reason to believe that vaccination has preserved thousands, who from a weak constitution would formerly have succumbed to small-pox. Thus the weak members of civilised societies propagate their kind."[27] Numerous eugenicists would echo that concern. In the post–World War II period, the medical advances most commonly cited as worrisome for the future were the development of powerful antibiotics that reduced deaths from infection, the 1922 invention of a method for isolating and purifying insulin, and beginning in the late 1950s, dietary treatment for the inherited metabolic disorders galactosemia and PKU. Thus, writing of the 1960s, bioethicist Albert Jonsen noted that "geneticists worried that the gene pool was becoming polluted because the early death of persons with certain genetic conditions was now preventable; in addition to antibiotics, insulin for diabetes and diet for phenylketonuria were frequently mentioned."[28]

Prior to the 1960s, the average life expectancy of individuals with PKU was short; only about 25 percent lived beyond the age of 30, with most deaths being the result of infection in institutions.[29] With the advent of dietary treatment, life expectancy could become normal. Of course, PKU was a rare disease. But as we saw in chapter 3, the ability to effect a "cure" through diet was viewed as a harbinger of successful control of numerous inherited diseases. For many geneticists, the salvaging of individuals who would ordinarily not have reproduced thus seemed a mixed blessing. Unless the germ line could be directly altered, the incidence of these diseases would probably slowly but surely creep upward. No geneticist advocated forsaking treatment. But many worried that, in the absence of measures to control reproduction by carriers, medical progress would be achieved at an ultimate cost to the gene pool.[30]

Følling, Jervis, and even Penrose, among other early researchers, hoped to develop methods of heterozygote detection so that carriers could be dissuaded either from reproducing or, as Penrose argued, from reproducing with each other. Before the development of screening and treatment for PKU, this seemed the only practical way to prevent the disease. But once screening and treatment were routinized, concern shifted to consequences for the gene pool, defined in the 1970 NAS report as "the primary resource of mankind, today and tomorrow." The committee explained that the health, longevity, and intellectual capacity of most humans today reflect the historical action of natural selection in minimizing the incidence of genes that, when homozygous, result in serious physical or mental impairment. But modern medical advances have relaxed the process of selection. Today, the report argued, we practice euphenics; that is, the engineering of human development in ways that permit the survival and reproduction of those with genetic handicaps, ultimately increasing the numbers of those affected. The report's primary example was dietary therapy for phenylketonuria, but the committee noted that its short list of treatable genetic disorders "must grow as clinical medicine learns to circumvent the consequences of many other genetic disorders."[31]

In a pair of speeches given at the dedication of the new Mt. Sinai College of Medicine in 1968, two Nobel laureates lent their scientific credibility to arguments about whether people with "genetic defects" should reproduce. British immunologist Sir Peter Medawar and American biochemist Linus Pauling, both politically progressive and eminent scientists, employed the case of PKU to argue the need for carriers of severely deleterious recessive genes to restrict their reproduction.

The nature of Medawar's argument may come as a surprise, for he is known as a critic of eugenics. On both scientific and moral grounds, he was indeed harshly critical of the kind of *positive* eugenics championed by Francis Galton, which aimed to enrich the population in a few superior genotypes. Thus, Medawar lambasted Muller's sperm-banking proposal, arguing that we do not know enough about the inheritance of mental and behavioral traits to engage in a breeding program and that the degree of heterozygosity in human populations implies that, even if we were able to agree on who belonged to the genetic elite, we could not count on it to breed true. Underlying efforts to breed such an elite, he charged, were "the morality of the gas chamber."[32]

Harsh words indeed. But Medawar also consistently contrasted positive and negative eugenics, with the latter said to have "the altogether lesser and more realistic ambition of diminishing, and as far as possible correcting, the distress caused by deleterious genes and genetic conjunctions." Even here, Medawar sometimes appeared as a critic, arguing that the sterilization of mentally deficient people would be ineffective, since most of the offending genes were hidden in symptomless heterozygotes, whom it would be folly to sterilize. He also noted that the relaxation of selection would only slowly increase the incidence of disease-causing genes and that genetic medicine of the future might succeed in repairing the damage.³³ But he nonetheless warned that if the sick "are to be kept alive and restored by medical procedures to something approaching a state of normal health, as it is right that they should be, then whatever elements of their genetic make-up may have contributed to their diseased state will for that reason be disseminated more widely throughout the population," and he argued that carriers of the gene for PKU should be dissuaded from marrying each other, since, on average, half their children would also be carriers and a quarter would have PKU. Alas, even if carriers complied, the problem would still not be solved, for if they married "normals," the result would still be a slow but steady increase in the frequency of the offending gene. "We are dealing here with a genetic equivalent of inflationary economics," he explained, for "we seem to be getting on all right, but the currency is deteriorating." He also asserted that "it is humbug to say that such a policy [discouraging PKU heterozygotes from marrying each other] violates an elementary right of human beings. No one has conferred upon human beings the right to produce maimed or biochemically crippled children."³⁴

The second main address at the 1968 dedication was delivered by Medawar's fellow Nobel laureate Linus Pauling. Although Pauling's achievement in the medical arena is primarily associated with sickle-cell anemia, whose molecular basis he identified in a celebrated 1949 paper, he had a long-standing interest in mental disease, which he considered the most significant source of suffering in the United States. In 1955, he received a large five-year grant from the Ford Foundation for work on the biochemical basis of mental deficiency. A goal of the project was development of an intravenous phenylalanine tolerance test to identify heterozygotes for PKU, who could then be warned not to marry other carriers.³⁵

Pauling's primary aim was to elucidate the mechanisms involved in the production of mental deficiency in phenylketonuria. Therapy for PKU, which involved limiting the supply of dietary phenylalanine in order to approximate normal blood concentrations of the amino acid and so alleviate the manifestations of the disease, was a paradigmatic example of his "orthomolecular" approach to mental disease.³⁶ Pauling had also long been concerned with the threat of genetic deterioration. Like Muller, by whom he was strongly influenced, Pauling believed that exposure to ionizing radiation, combined with advances in medicine, was dangerously increasing the human mutation rate. "The human race is deteriorating," he had warned in 1959. "We need to do something about it."³⁷ Unlike Muller, whose anticommunism was even stronger than his fear of genetic damage from radiation, Pauling vigorously campaigned to end atmospheric nuclear testing—leading to his winning a second Nobel Prize (the Peace Prize). He also departed from Muller on the desirability of positive eugenics. Pauling feared that a program of "germinal choice" would ultimately result in fewer contemplative intellectuals and more social conformists. On the other hand, like Medawar, he considered negative eugenics eminently sensible, since "no objection can be legitimately raised . . . against the ambition to eliminate from human heredity those genes that lead to clearly pathological manifestations and great human suffering."³⁸

In Pauling's view, couples with an already affected child should have no further children, since, even if their offspring were normal, they would be carriers of genes that cause suffering. Thus the law should require the testing of all individuals in communities where the incidence of genes for particular molecular diseases is high, with carriers then refraining from marrying each other or having biological children.³⁹ In a 1968 essay, Pauling famously proposed that legislation be adopted to require young persons to have tattooed on their foreheads symbols for any seriously defective recessive genes, such as those for sickle-cell anemia and PKU, so that they would refrain from falling in love with each other—a proposal he reiterated at Mt. Sinai.⁴⁰ Like Medawar, he realized that such a "masking" strategy would do nothing to decrease the incidence of the defective gene; he thus suggested that carriers who married noncarriers should have a lower than average number of offspring.

The speeches at Mt. Sinai were no aberration, at least among the group of scientists who pronounced on social and ethical issues in genetics. Indeed, proposals to limit reproductive freedom were so widely discussed in the 1960s that a writer for *Harper's Magazine*, reporting on a proposal to issue licenses to reproduce only to those whose genes received a passing grade, commented that "eugenic proposals like this are commonplace at scientific meetings nowadays. After twenty years of ill repute, eugenics is again the subject of respectable scientific investigation."[41]

"GOOD" AND "BAD" EUGENICS

Of course, the morality and utility of eugenics were also vehemently criticized, and there were important countercurrents. Proponents of schemes for selective breeding often complained that they were swimming against the tide, especially if they advocated compulsion. Oppositional forces were building and would soon swamp efforts to rehabilitate both the term "eugenics" and *any* principle of reproductive responsibility. As noted earlier, in 1967, 30 nations had signed the UN Universal Declaration of Human Rights, a statement that described the family as "the natural and fundamental unit of society" and proclaimed, "It follows that any choice and decision with regard to the size of the family must irrevocably rest with the family itself, and cannot be made by anyone else." In the same year, Kingsley Davis, who served as president both of the American Sociological Society and of the Population Association of America, complained that "in the sphere of reproduction, complete individual initiative is generally favored even by those liberal intellectuals who, in other spheres, most favor economic and social planning."[42] Thus eugenics had already been stigmatized in some professional communities, including most of the social sciences. But in others, including genetics, there was surprisingly little reluctance to characterize favored policies as "eugenic." Rather, as Edmund Ramsden notes, geneticists typically tried to transfer stigma to an earlier generation of eugenicists.[43] The bad, scientifically simplistic, racist, and class-based eugenics of the past was frequently counterpoised to the good, scientifically sound, and racially and socially unbiased eugenics of the present. Moreover, even those geneticists who would have rejected the eugenicist label assumed that it was wrong to knowingly transmit a serious genetic disease.

Given the number of discussions in scientific circles about the need for reproductive restraint, it is understandable that many commentators at the time thought they had spotted a trend. For example, Harvard historian Donald Fleming wrote in 1969, "What we may reasonably expect is a continually rising chorus by the biologists, moralists, and social philosophers of the next generation to the effect that nobody has a right to have children, and still less the right to determine on personal grounds how many."[44] Similarly, Ramsey thought it could "safely be predicted that the future will see more rather than less discussion of proposals for genetic control." And medical geneticist Joseph Dancis remarked in 1973 on the growing sentiment "among both physicians and the general public that we must be concerned not simply with the birth of a baby, but of one who will not be a liability to society, to its parents, and to itself." The next year, the author of a *Fortune* article commented that "a good many geneticists hold that society ought to encourage parents to avoid giving birth to children crippled by such genetic ailments as mongolism." The author cited Cal Tech biologist James Bonner, who anticipated "the rise of 'a new morality' in which couples will say, 'Since we will have only two children, let them be free from genetic defects.'"[45]

As it turns out, these predictions proved utterly wrong, at least regarding public statements about eugenics. It is true that individual families have used prenatal diagnosis to avoid having a child with a chromosomal abnormality. But ideas about society imposing sanctions on couples to follow eugenic principles would, by the 1990s, seem hopelessly misguided. Whether a genuine shift in public opinion occurred is impossible to say. But the opinions of those "authorized to speak for" the public were certainly transformed.[46] The decline of the discourse about reproductive responsibility is attributable to the emergence of social movements with very different ideological foundations: the closely linked patients' rights and feminist movements, for which respect for autonomy is a core value, and the disability rights movement, for which reproductive choices that are autonomous might still be morally wrong and should be discouraged even if not prohibited by law.

THE DISABILITY RIGHTS MOVEMENT

In the 1960s, when so many prominent scientists and theologians were asserting the immorality of knowingly giving birth to a "defective" child, a movement was building that rejected the assumption that a person could be inherently burdensome. The disability rights movement championed a redesign of institutions that turned cognitive and physical impairments into social disabilities, as well as promoting an attitudinal change such that individuals would no longer be equated with their impairments or viewed as objects of pity. The movement would score several major victories. In the 1970s, children with disabilities had no right to a public education, and adults were routinely excluded from the workplace, public places, and private businesses. Within a generation, disability rights advocates transformed laws, institutions, and public perceptions, and federal laws in the United States required the provision of an appropriate public education in the least restrictive environment for every child and appropriate accommodations in public places for both adults and children. In Australia and the United Kingdom, decades of activism also resulted in passage of national legislation barring discrimination against people with disabilities in employment, education, and the provision of goods, services, and facilities.[47] By the mid-1990s, discrimination in many countries not only had become illegal but was perceived as morally wrong.

With respect to cognitive impairment, a key event in this history was the publication, in 1961, of Canadian sociologist Erving Goffman's *Asylums*, a collection of essays that described how the daily life and unwritten rules of institutions shape behavior, mostly in negative ways.[48] Danish reformer Niels Erik Bank-Mikkelsen sought to put Goffman's ideas about institutions into practice in his role as director of the Danish Service for the Mentally Retarded. Based on his experience in a Nazi concentration camp, Bank-Mikkelsen argued that the conditions of institutions for people with MR were dehumanizing and that community-based services with full civil rights were not only morally necessary but therapeutic. Bank-Mikkelsen is described as the father of "normalization," in which a way of living that follows as closely as possible the patterns and conditions of everyday life is the goal for people with MR. Swedish reformer Bengt Nirje further developed and popularized the concept of normalization, and in North America, Wolf Wolfensberger applied the principle to all socially disadvantaged groups.[49]

By the 1970s, normalization became the centerpiece of the disability rights movement, as activists sought full acceptance as equal citizens rather than being viewed as diseased individuals to be pitied or cured. To achieve their goals, disability advocates followed the examples set by the Civil Rights and feminist movements. For example, even though a series of federal laws and landmark court cases in the 1970s made discrimination in public institutions illegal, advocates resorted to public demonstrations and occupation of government offices to ensure enforcement of the law.

Their basic premise was that although people may have a variety of physical or developmental impairments, it is society's failure to make accommodations that produces disability. Advocates of this "social model" of disability argued that with appropriate supports, most individuals with developmental disabilities could live independently. While the dominant medical model sought to "fix" people with disabilities, the social model called for curb cuts and wheelchair ramps to help a child using a wheelchair access a public school education, for example, and elevators with numbers in Braille and audible crosswalk signals to assist independent mobility for a person with a visual impairment.[50]

Disability rights activists also challenged the view that it would be better for everyone if some kinds of people were not born. In the 1970s, children born with Down syndrome were routinely institutionalized or allowed to die from conditions that would be treated in nonimpaired infants. Disability rights advocates argued that the suffering ostensibly endured by individuals and their families because of a disability was typically exaggerated and the quality of life underestimated, especially by scientists and clinicians. They also insisted that all life was equally worthy, a perspective that led to unease with prenatal diagnosis and, later, preimplantation genetic diagnosis, which inevitably involve discriminatory judgments.[51] Many Catholic and conservative opponents of abortion also view the use of prenatal and preimplantation diagnosis as a reflection of mistaken ideas about and callous attitudes toward people with disabilities.[52] This shared viewpoint was codified into an amendment of the federal child abuse laws in 1984, after US Surgeon General C. Everett Koop publicly criticized a surgeon for failing to operate on a baby merely because the infant had Down syndrome. The law requires states that receive federal funds for combating child abuse to develop reporting procedures designed to prohibit the withholding of medical care from infants, except when a baby is irreversibly comatose or the treatment is "virtually futile."[53] The so-called Baby Doe regulations have had a substantial impact on how neonatologists consider end-of-life decisions for newborns.[54]

FEMINISM, BIOETHICS, AND THE TRIUMPH OF AUTONOMY

The disability rights challenge to the prevailing scientific and medical discourse on disability is ideologically distinct from and, indeed, sometimes in tension with the autonomy-based feminist movement, emerging in the same period. Both developments, in their different ways, undermined the view that there is a social responsibility in reproduction that is violated by risking the birth of a child with a serious impairment. Growing acceptance of the principle of respect for autonomy is probably the key reason that the tone of public discussion shifted away from eugenic ideas.

The patients' rights movement and, especially, the feminist movement of the 1960s followed on the Civil Rights and anti–Vietnam War campaigns and included many of the same protagonists. In denouncing doctors' authority as patriarchal, the movements adopted "choice," "autonomy," and "self-determination" as their slogans.[55] The title—and popularity—of the 1971 book *Our Bodies, Ourselves* neatly captured how autonomy implied "control over our bodies, our labor and economic resources, our life decisions" and, as such, was viewed by feminists as central to the achievement of their political goals.[56]

At the same time, reproductive genetic services began a period of rapid expansion. Prior to the 1970s, few individuals made use of the limited genetic counseling services then available. Most of those who did came from families with a history of a particular disorder or already had an affected child. All the counselor could offer was an (often imprecise) estimate of risk. Moreover, clients' choices were severely limited, since the risk could be avoided only by refraining from childbearing. Under these conditions, genetic counseling had little to offer, and it remained a small-scale affair, mostly practiced by PhD geneticists and by physicians trained in genetics.[57] Before 1970, these clinicians generally believed that children with MR should be institutionalized for the good of the family and that the births of further children with MR should be prevented for the good of the race.[58]

Genetic counseling acquired greater practical importance in the 1970s, when the development of prenatal diagnosis coincided with the legalization of abortion. Amniocentesis was first developed in the 1960s and entered clinical practice in the 1970s. In Britain, abortion was legalized by an Act of Parliament in 1967; in the United States, several states relaxed restrictions in the 1960s, and the US Supreme Court prohibited states from unconditionally barring the procedure in the 1973 case of *Roe v. Wade*. As a result of this convergence, the demand for genetic services—and hence counselors—exploded. The earliest professional master's degree program in genetic counseling, at Sarah Lawrence College, graduated its first class in 1971. Its students were trained in techniques of Rogerian therapy, according to which the role of professionals was to clarify the client's own values, not impose their own. The education of these overwhelmingly female students also coincided with the rise of the autonomy-oriented feminist movement. At Sarah Lawrence and in other programs that quickly followed in its wake, respect for patients' autonomy—often understood to imply a stance of "nondirectiveness"—became a bedrock principle.[59]

It is of particular importance that genetic services expanded in the context of impassioned controversy over the morality of abortion. That charged social context guaranteed that medical geneticists and genetic counselors would stress clients' freedom to make their own decisions. After all, these professionals did not want to be accused of fostering such a contentious practice as abortion. Given the paucity of treatments for genetic conditions, the accusation that prenatal diagnostic services promoted abortion had considerable force. Denying that there was any single, correct reproductive decision functioned to defuse this charge—hence the insistence that the goal of reproductive genetic services, or at least the only acceptable goal, is increasing the choices available to women. According to the official view, "autonomous decision making should be the goal in prenatal diagnosis and . . . health professionals, society, and the state [should] be neutral on the outcome of individual reproductive choices. Reproductive genetic services should be aimed at increasing individual control over reproductive options and should not be used to pursue eugenic goals."[60]

Also germane to the apparent shift in ethos is the emergence of bioethics as a distinct intellectual discipline, whose practitioners would become the new arbiters on a wide range of ethical issues in medicine. In 1979, Thomas Beauchamp and James Childress published their influential *Principles of Biomedical Ethics*.[61] Although they originally proposed autonomy as one of four equally important values in bioethics—along with beneficence, nonmaleficence, and justice—autonomy quickly became the key principle guiding clinical and research ethics.[62] Respect for autonomy has frequently been interpreted as implying that a person "should be free to perform whatever action he wishes—even if it involves serious risk for the agent and even if others consider it to be foolish."[63]

The reasons for the emphasis on autonomy are complex. It is

certainly significant that bioethics developed in the context of a series of scandals involving experiments on human subjects, most notably the thalidomide disaster of 1961, when testimony revealed that drug companies provided doctors with samples of experimental drugs, then paid doctors to collect data on their patients; Henry Beecher's 1966 exposé of risky experiments conducted on patients at distinguished medical institutions, without the patients' knowledge or consent; and the 1972 exposé of the syphilis experiments at Tuskegee.[64] These and other scandals undermined the assumption that physicians could be trusted to act in their patients' best interests and therefore strengthened the case for patients' autonomy. Given its emergence in the 1970s, bioethics was also inevitably affected by the other emergent social movements, including Civil Rights and feminism, and by the general cultural fracturing of the time. As Albert Jonsen notes in his history of the field, "In a pluralistic society, where broad agreement on the content of morality seemed to be fading, a principle of autonomy, as the sole or primary moral principle, solved many a conundrum; one merely respects the wishes and choices of every person without passing judgment on further moral grounds."[65]

The principle of respect for autonomy came to dominate bioethics at a time when bioethicists began to replace scientists as the primary spokespersons on social and ethical issues in genetics. During the 1950s and 1960s, books and articles on social issues in genetics were primarily authored by scientists. And it was scientists to whom journalists, foundations, and conference organizers typically turned for guidance on genetics-related ethical and social issues. Indeed, at the four major symposia on this theme held in the 1960s, nearly all the participants were scientists; the debates were primarily between geneticists who believed that genetic engineering of humans was premature and those who believed that its time had come.[66] But by the 1970s, bioethicists had begun to influence the discourse on issues in genetics.

Reproductive counseling in the case of PKU reflects these larger cultural shifts. Almost no one has worried, for several decades, about the prospect of genetic deterioration resulting from the reproduction of carriers for PKU or other recessive diseases. In the early 1980s, lingering concerns about the long-term consequences of treatment for PKU were largely displaced by new concerns about damage from untreated pregnancies in women with the disease, a threat of much greater urgency. But by the time the magnitude of the problem of maternal PKU was recognized, the principle of reproductive responsibility was already becoming unfashionable. Eugenic concerns—whether related to the future quality of the gene pool or to the next generation—were increasingly muted. Direct advice giving came to violate the cultural norms both reflected in and fostered by the new field of genetic counseling.

The rise of autonomy is illustrated by the results of a survey conducted in the mid-1990s that found 59 percent of geneticists and 39 percent of primary care physicians would support a parental decision to give birth to a child even when maternal PKU carried a 100 percent chance of birth defects.[67] Those statistics reflect not only an increasing acceptance of the principle of respect for autonomy but significant differences related to field of training. In the same survey, only 10 percent of geneticists but 78 percent of physicians and 81 percent of patients agreed that people at high risk for transmitting a serious genetic disorder should not have children unless they used prenatal diagnosis and selective abortion. That is, geneticists were much more permissive in allowing families to make reproductive decisions, even if these decisions resulted in a child with a disability or chronic medical condition. In general, physicians and the general public seemed to think it unfair to the child, to siblings, and to society to knowingly run such a risk.[68]

In the 1960s, 1970s, and even 1980s, the practices of physicians, psychologists, nutritionists, and other clinicians in PKU clinics were more directive than the information-providing style that predominated among medical geneticists and genetic counselors. Indeed, women with PKU were rarely referred to professional genetic counselors, given that the issue was not the easily calculable chance that the child would have PKU but rather the nongenetic problem of maternal PKU. The rise of patients' autonomy in clinical ethics and the role of the feminist and disability rights movements together have changed the calculus of reproductive decision making. The penetration of the genetics model has been so complete that today's metabolic clinicians are often incredulous to learn that, not so long ago, identifying PKU carriers was considered an important goal and advising young women with PKU not to have children was a widely accepted practice.

CHAPTER TEN

Newborn Screening Expands

❖ ❖ ❖

Newborn screening is more than the collection of blood from newborns to detect an unseen malady. It is a complex system of care that includes training of personnel in hospitals throughout the country; collection of individually identified blood spots and their transfer from birthing hospitals to a centralized laboratory; high-volume testing for rare conditions at state-sponsored labs; continuous quality assurance for sophisticated tests, using national standards; notification of families, physicians, and hospitals about presumptive positive results for confirmatory testing; timely treatment of newborns by specialists in metabolic, endocrine, and hematological conditions; and a legislative and regulatory superstructure that varies by state. As noted in previous chapters, the history of PKU is unusual, because it is a rare condition yet serves as a paradigm for several aspects of modern medicine, particularly genetics. Beyond its symbolic meaning, the history of newborn screening for PKU is critical because it provides the framework for the remarkable expansion of NBS in the early twenty-first century.

THE EARLY EXPANSION OF NEWBORN SCREENING

After his success with newborn screening for PKU, Robert Guthrie continued to develop new bacterial assays for metabolic disorders that employed the same principle of competitive inhibition between an amino acid and one of its chemical analogues. In the 1960s and 1970s, researchers in his laboratory and elsewhere also developed blood-spot screening and confirmatory tests utilizing enzyme analysis, fluorometry, and chromatography, as well as tests for other types of inherited disorders.[1] The hope was that PKU would be the first of many metabolic diseases amenable to early treatment and that the addition of other conditions would eventually reduce the cumulative burden of mental retardation. Edwin Naylor, who would play a key role in the expansion of NBS in the 1990s, began a postdoctoral fellowship with Guthrie in 1972. He recalls that the lab was then focused primarily on "developing tests for other conditions that could be piggybacked onto the blood spot PKU assay," noting that "once you have the original system in place where you were collecting the blood spot, it was sort of a shame just to be doing PKU when you could use that blood spot for screening for other preventable or treatable disorders."[2] By the mid-1970s, as many as 11 conditions could be detected with the same filter-paper blood specimen used to test for PKU; the most common additions were maple syrup urine disease (MSUD), homocystinuria, galactosemia, hereditary tyrosinemia, cystinuria, and histidinemia.

A technological advance was crucial to this initial round of NBS expansion. Guthrie had always been keenly interested in ways to make screening less labor- and skill-intensive. The kits he had assembled for the US Children's Bureau (USCB) field trial of the assay for PKU were intended to make the procedure as simple as possible; that the kits could be used by technicians with no training in bacteriology was fundamental to their success. In 1964, California inventor Robert Phillips, a sometime collaborator of Guthrie's, developed a machine to automate the preparation of newborn blood spots for analysis. The Phillips Punch Index Machine, which replaced the single hand-held paper punch, allowed four specimens to be punched simultaneously from a single blood spot and distributed mechanically into four separate trays.[3] Recalling his early days in a NBS lab, Bradford Therrell described how "you punched the lever, it punched four holes. They were held by vacuum on four little arms, and these arms would disperse and drop the specimen into four index locations at the same time. Row one, column one, on four different media for instance. So doing it that way, with the same energy to punch one, you could punch four tests . . . So then it just became a game of finding the different types of systems that would fit into this quadruplex machine . . . laboratories began to go to four tests, or then they would go to eight tests if you punched everything twice. So that became the thing, that you first expand to four, and then if you go beyond four, you might as well do eight."[4]

One powerful incentive to add new tests was the need to demonstrate the value of NBS, including its cost-efficiency. In the early years of screening for PKU, newspapers and popular magazines ran stories highlighting problems with diagnosis and treatment, often quoting critical statements by Samuel Bessman or the American Academy of Pediatrics.[5] Furthermore, the low incidence of PKU and the fragmented nature of the screening system in the United States meant that relatively few cases of the disease were identified in the early years of NBS. The small number of cases, combined with unexpected problems that emerged in the mid-1960s, prompted widespread reappraisal of the value of screening programs. Thus, according to a 1971 article in the *Medical World News*, "One of the biggest questions being mulled now is also one of the most fundamental: Is mass screening really worthwhile?" Citing Rudy Hormuth of the USCB, the article noted that most states were taking a "hard look" at their PKU testing programs and that experts were "asking if we can afford to spend $3 to $4 apiece to test each child born in this country just to detect 200 infants a year with classic PKU and another 100 with hyperphenylalaninemia."[6] Indeed, in 1970, after three years of screening its largely black population without detecting a single case, Washington, DC, repealed its screening law, concluding that scarce public health resources would be better used elsewhere. Several other jurisdictions threatened to follow suit.[7]

Guthrie responded to such critiques by insisting that, when all

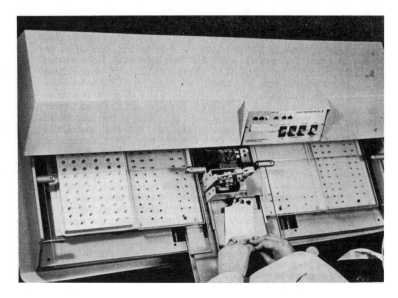

Phillips Punch Index Machine with tray.
Courtesy of Oregon State Board of Health.

relevant factors such as lost productivity were considered, screening *did* save states money and could be made even more cost-efficient. Regionalization, which Guthrie strongly promoted, was one means to accomplish that end; so was adding new tests to the existing system. After all, more blood was collected than was needed to detect PKU, and the system for collecting the specimens, forwarding them to the laboratory, and reporting the results was already in place. In an automated system, adding new tests would involve only marginal costs. Guthrie and other supporters of newborn screening deeply believed that expanding screening would benefit children, their families, and the larger society. In his view, not to add inexpensive tests such as those for MSUD, galactosemia, and homocystinuria would be "a tragic waste of an opportunity to prevent serious disease in children."[8]

Despite Guthrie's efforts, expansion was slow. Until the mid-1970s, PKU was the only condition for which most jurisdictions, both in the United States and internationally, screened newborns.

That situation changed when Jean Dussault and colleagues in Quebec discovered that they could detect congenital hypothyroidism (CH), a thyroid hormone deficiency, by carrying out a radioimmunoassay for thyroxin on the Guthrie card samples. CH is about five times more common than PKU, with an incidence of 1 in 3,000 births, and because it is usually sporadic and is not genetic, the incidence does not vary significantly among different populations.[9]

CH also had other characteristics that made it an attractive candidate for screening. It was difficult to diagnose clinically, had drastic effects on cognitive development that were preventable with early detection and treatment within three months of birth, and could be effectively and inexpensively treated with a daily oral dose of synthetic L-thyroxin.[10] In 1974, the province of Quebec, which had been screening for PKU and hereditary tyrosinemia, became the first jurisdiction to screen for CH. Many US states and several countries quickly followed, making NBS much more cost-effective.[11] Indeed, the availability of CH testing was the main reason that the District of Columbia reinstated its NBS program.[12]

In 1973, Michael Garrick, then a research associate in Guthrie's lab, adapted a type of electrophoresis to detect sickle-cell disease (SCD) and other hemoglobinopathies, using dried blood spots on filter paper.[13] With the appearance of this technique—a year after passage of the National Sickle-Cell Anemia Control Act and in the context of intense interest in the disease and the passage of many state laws promoting or mandating sickle-cell screening of schoolchildren and marriage license applicants—a few states also began to test newborns for SCD. But there was no effective treatment for the disease, and uptake of the test was limited.[14] Then, in the late 1980s, evidence that penicillin prophylaxis in infants with SCD was effective in reducing mortality and morbidity prompted widespread hemoglobin electrophoresis testing for the disease. By 1990, 16 states had adopted universal screening, with 14 others screening high-risk populations or with programs in development.[15]

THE 1990S: NEW TECHNOLOGY, NEW DISPARITIES

Screening coverage in the United States was highly uneven by the end of the twentieth century, with some jurisdictions testing for only a few conditions—most often, PKU, CH, galactosemia, and SCD—and others testing for many. Both the pace of expansion and the coverage disparities increased dramatically in the 1990s with the introduction of tandem mass spectrometry (MS-MS), a technology that measures the molecular mass of molecules. In the early 1990s, David Millington and colleagues at Duke University applied MS-MS to the quantitative analysis of amino acids and acylcarnitines (fatty and organic acids bound to the chemical carnitine) in dried blood spots and demonstrated that it could be used to screen for PKU and several other disorders.[16] The advent of MS-MS hugely increased the number of conditions detectable in newborns; indeed, the technology was capable of measuring hundreds of disease-associated metabolites in a single rapid assay, as well as providing more sensitive and specific screening for PKU and several other conditions already included in many state NBS panels. With Guthrie testing, each condition added to the NBS battery required development of a new test specific to a disease product. This, along with the limit on the amount of blood that could be taken from an infant, was a significant constraint on test proliferation. MS-MS is a single process that can be used to identify hundreds of disease products at once and does not require blood samples any larger than those already being collected. Apart from the initial outlay for the machine and for technician training, MS-MS required little more in either time or money to test for many than for a few conditions.

With the availability of MS-MS, many jurisdictions began to expand their screening panels to include additional amino acid disorders and also, given the ability to detect abnormal levels of acylcarnitines, disorders of fatty acid oxidation and organic acid degradation. In 1995, US states mandated screening for an average of 5 conditions; during the next decade, they added another 19, on average, with several adding more than 40.[17] The expansion was accompanied by increasing disparities among states, such that by 2005, the number of conditions included by law or rule in states' screening panels ranged from 3 to 43.[18] This disparity occurred because, unlike most countries that established screening programs in the 1960s and 1970s, the United States had no national NBS policy.[19] Expansion was haphazard and unpredictable: the 1975 National Academy of Sciences report on genetic screening noted that state and local medical societies had little if any involvement in decision making and alleged that, often, a test would be added "simply because it can be done by a modification of the Guthrie test and uses the sample taken for the PKU screening."[20]

The widening differences among states after 2000 prompted calls for standardization to ensure that no infants suffered just because they were born in a state that did not screen for a condition—and to avoid lawsuits by parents seeking redress when their infant was harmed by lack of screening. In this context, the American College of Medical Genetics (ACMG) was asked to evaluate NBS and develop recommendations for a uniform condition panel.[21] Its report, released in 2005, strongly endorsed the use of MS-MS and other new multiplex technologies that allow near-simultaneous screening for multiple conditions from a single specimen. The scoring system employed by the ACMG committee awarded 200 points to a candidate condition if it could be detected using a multiplex platform.[22] The report also recommended that MS-MS be operated in "full-profile" mode rather than operated to selectively target the conditions deemed appropriate for screening.[23] Both the scoring system and the recommendation to open up the whole MS-MS spectrum to screening would arouse intense controversy, as would the seeming lack of analysis of potential costs and harms of screening in the ACMG report.[24]

Eighty-four candidate conditions were evaluated and categorized as "core," "secondary," or inappropriate for NBS. Conditions identified as core or primary targets for screening were considered to have a sufficiently well-understood natural history, a test with appropriate sensitivity and specificity, and an available and efficacious treatment.[25] Conditions were categorized as secondary if they fell short of that standard—for example, lacked an effective treatment—but could be identified in the course of screening for the core panel, had received a score in the moderate range, and were thought to merit inclusion on other grounds. The committee recommended that all states screen for a core panel of 29 conditions and report on an additional 25 secondary conditions. It also recommended that carrier status, when definitively (though incidentally) identified, be communicated to health care professionals and families.[26]

The report's recommendations were quickly accepted by numerous bodies, including the American Academy of Pediatrics, a host of disease-specific advocacy groups, and the March of Dimes, a late but influential convert to the NBS cause. The most important endorsement came from the Secretary's Advisory Committee on Heritable Disorders in Newborns and Children (SACHDNC), which advises the secretary of the US Department of Health and Human Services on screening policy and greatly influences federal agencies, state governments, and health insurers. The chair of SACHDNC, R. Rodney Howell, had also served as a key author of the ACMG report. With SACHDNC's and then the Health and Human Services secretary's approval, uptake of the report's recommendations was rapid. Within a few years, every state was screening for all 29 core conditions, and the majority for at least 20 of the secondary conditions. As of 2012, SACHDNC recommended that every newborn screening program should screen for 31 core and 26 secondary conditions, and most states are heeding its advice.[27]

TESTING AT THE BEDSIDE

Newborn screening in the United States has also recently expanded to include conditions such as hearing impairment and congenital heart disease that involve clinical tests at the bedside rather than analysis of blood in a centralized laboratory. Endorsed by many professional groups in the 1990s, newborn hearing screening is now virtually universal.[28] Screening for congenital heart disease was recommended by SACHDNC in 2010, was endorsed by the secretary for Health and Human Services in 2011, and has been implemented by at least 10 states.[29] In both cases, relatively simple bedside tests can detect potential difficulties, resulting in better outcomes for infants.

Why are these tests integrated into state NBS programs, given that they are not fundamentally different from other routine newborn clinical tests, such as blood glucose monitoring or newborn hip exams, which are performed in all neonates but not included in state NBS programs? A primary reason is to ensure that all newborns are tested and have appropriate follow-up. When incorporated into state NBS programs, screening for hearing impairment and congenital heart disease are subject to some sort of reporting requirement, to judge whether newborns are being screened and those affected are being treated. In private medicine, the quality of routine clinical care depends much more on the vagaries of clinician training, the capabilities of a particular institution, and such social factors as the availability of health insurance. In earlier chapters, we saw that the system of care for NBS is far from perfect. Nevertheless, a striking feature of the history of NBS is the involvement of the state in ensuring universal quality and access as part of a public health mandate—an anomaly in the American health care system, where oversight is often weak and equity lacking.

SCREENING INTERNATIONALLY

Guthrie's biographer noted that he "traveled to the corners of the earth in his tireless efforts to inspire converts."[30] That is no exaggeration. In the decade following establishment of the first US programs to screen for PKU, Guthrie made frequent and extensive trips abroad, including a nine-month around-the-world trip in 1965, during which he forged connections both with parents' groups and with national and international professional societies. Apart from several European trips, Guthrie's campaign for bloodspot screening took him to Australia and twice to New Zealand (including for a sabbatical year in 1968–69), to several islands in the South Pacific, and to South Africa, Tokyo, India, Iran, and Israel. Most of these trips were funded in whole or in part by the USCB, which also sponsored international meetings in Yugoslavia, Germany, Egypt, Israel, and Poland. The first international conference, held in Dubrovnik in 1966, included participants from Poland, Belgium, Germany, Ireland, Sweden, Scotland, Greece, India, Israel, Mexico, and Pakistan, as well as Yugoslavia; there were only two Americans at the conference, including Guthrie.[31]

By the end of the 1960s, blood-spot programs to screen for PKU had been introduced in numerous countries.[32] Canada's first program was launched in the province of Prince Edward Island in 1963, and with the exception of Newfoundland and Labrador, all other provinces and territories had developed programs by the end of the decade.[33] Guthrie's extended 1965 trip resulted in one of the first national blood-spot screening programs when New Zealand adopted a countrywide program for four conditions—PKU, galactosemia, MSUD, and hereditary tyrosinemia—in 1966.[34] Blood-spot screening for PKU began in Australia in 1967–68 and by the end of the decade had been implemented state- or territory-wide.[35] Screening programs were also instituted in Japan (as a pilot program) in 1967, in Israel in 1966, and by 1970 in much of Western Europe.[36]

Perhaps more surprisingly, several programs were also established in relatively poor countries where one might not expect screening for rare disorders to be a health priority. A crucial factor in the dissemination of the US program to countries like Pakistan and Poland was the USCB's use of funds made available from the sale of American surplus farm commodities. Under the US Agricultural Trade and Development Assistance Act of 1954, also known as "Food for Peace" or Public Law 480 (PL-480), foreign currencies acquired from sale of these surplus foodstuffs could be accessed by federal agencies for medical and scientific research and other health-related activities in developing countries. The Children's Bureau used its PL-480 funds both to sponsor confer-

Robert Guthrie, next to map with pointer to Kuwait.
Courtesy of the University Archives, State University of New York at Buffalo.

ences abroad and to assist in the establishment of NBS programs. In the context of controversy over the cost-effectiveness of screening and suggestions that efficiency would be increased if particular populations were targeted, the decision was made to establish screening programs in foreign countries where data could be obtained on variations in the incidence of PKU among ethnic/racial groups.

Bureau staff did not anticipate an enthusiastic response to this effort, given the extent of unmet maternal and child health needs and scant possibility of treatment for PKU in the countries where PL-480 funds were available. They were thus surprised to find researchers in those countries eager to cooperate and suggested this response arose "partly from their desire to be associated with the West in something that is new and exciting, and partly from their realization that this program gives them an opportunity to develop laboratory and clinical facilities which can be used for a much broader program than the detection of one rare inborn error of metabolism." Moreover, such projects gave scientific workers "the sense of belonging to a modern scientific community." In the view of USCB staff, it was reasonable to take into account these workers' sense of isolation and their "feeling of being passed over by the march of science" in determining program priorities.[37] Within a year of establishment of the first blood-spot screening programs for PKU in the United States, programs had been started with USCB assistance in Poland, Yugoslavia, Israel, and Pakistan.[38] Programs in China, Japan, and Egypt would follow.[39]

Despite the influence of the United States on early NBS programs, most countries did not follow the US trajectory of expanding newborn screening. For example, until 2006, most Canadian provinces screened only for PKU and CH.[40] In the United Kingdom as well, only the test for CH was introduced nationally in the 30 years following the advent of blood-based screening for PKU. In Ireland, Germany, France, and several other countries in Europe, as well as in Israel and Japan, the initial rapid uptake of newborn blood-spot screening would be followed by a long period of relative quiescence.

The pace of expansion did quicken in some nations with the invention of MS-MS and its application to newborn screening, a trend intensified by publicity surrounding the commissioning of the ACMG committee and publication of its report. In 2005, Germany expanded its panel to 15 conditions. In 2007, the Netherlands extended screening from PKU, CH, and congenital adrenal hyperplasia to a total of 17 conditions, as did Israel, which until 2008 had screened only for PKU and CH. In Australia, the state of New South Wales adopted MS-MS in 1998, making it among the first jurisdictions in the world to do so, and by 2012, all NBS programs in that country screened for at least 30 conditions.[41]

This round of expansion was extremely uneven. For example, in 2012, Austria screened for 31 conditions and Hungary for 30, but Ireland, one of the first countries with a national screening program, screened for only 6.[42] In the United Kingdom, screening remained limited to 5 conditions: PKU, CH, SCD, cystic fibrosis, and medium-chain acyl-CoA dehydrogenase deficiency (MCADD).[43] Poland screened for 3 condition, while France had universal screening for 5—PKU, CH, congenital adrenal hyperplasia, cystic fibrosis, and MCADD—along with SCD in targeted populations. In Scandinavia, Finland screened for 1 (CH), Norway for 2, Sweden for 5, Denmark for 15, and Iceland for 28.[44]

Generalizing across Europe is complicated by the fact that some countries, such as Italy and Spain, organize screening by region or province, without coordination at the national level.[45] But a survey sponsored by the European Union, completed in 2011, shows both that expansion has been accompanied by increasing disparities among countries and that its scope has generally been modest compared with the United States. Among the included 40 European countries—EU member states, candidate or potential candidate member states, and European Free Trade Association countries—screened conditions per country varied from 1 to 29. Only 5 of the countries surveyed—Austria, Hungary, Iceland, Portugal, and Spain—screened for more than 20 conditions, whereas 25 countries screened for no more than 5, and 21 countries for 3 or fewer.[46] The overall picture resembles that of the United States prior to publication of the ACMG report and has prompted the same calls for harmonization. Thus Jean-Louis Dhondt, a key figure in the French program, asked in 2010, "how does one explain to the parents of a child with brain damage caused by one of the screenable diseases that, if the child had been born in an adjoining country, he or she would have been screened and treated for the disorder and would be developing normally?"[47]

SCREENING CRITERIA

These disparities developed even though most NBS programs ostensibly applied the same basic criteria to the evaluation of new, candidate screening tests. In the mid-1960s, as medical screening programs for diabetes, hypertension, various cancers, infectious diseases, and other conditions proliferated, the World Health Organization commissioned an analysis of screening principles and practices. The result was an influential report, published in 1968, that articulated 10 criteria for the assessment of screening tests. (However, by then, most US states had already enacted screening statutes.) Usually referred to, after their authors, as the Wilson and Jungner principles, they have served as the starting point for most other efforts to define screening criteria.[48]

Wilson and Jungner's central idea was that while screening may seem an eminently sensible way to combat disease, the logic of early detection, when applied in the real world, is anything but simple: effective treatment does not always follow detection of real disease, and harm can occur to those who do not need treatment. They considered the need for criteria to guide the choice of conditions for screening "especially important when case-finding is carried out by a public health agency, where the pitfalls may be more numerous than when screening is performed by a personal physician."[49] In the years since then, other organizations and groups have articulated their own criteria, but they have generally retained the key tenets of Wilson and Jungner, especially the availability of an effective treatment. "Of all the criteria that a screening test should fulfil," Wilson and Jungner argued, "the ability to treat the condition adequately, when discovered, is perhaps the most important."[50]

Wilson and Jungner had explicitly endorsed screening for PKU despite its rarity, given "the very serious consequences if not discovered and treated very early in life." Indeed, PKU has remained the model for newborn screening, despite the difficulties in demonstrating treatment efficacy and persistent questions regarding whom to treat, what phenylalanine levels are safe, and when treatment should stop.[51] PKU seemed straightforward compared with conditions such as Duchenne muscular dystrophy, a lethal condition with no known treatment that has been detectable by NBS dried blood spot since 1975.[52] Most NBS programs worldwide have taken as axiomatic that screening is only defensible when it provides a demonstrable medical benefit to the infant, and as a consequence, conditions like Duchenne's have generally not been added to NBS programs.[53] To be sure, there have been generally acknowledged missteps in newborn screening programs, such as the addition of histidinemia in New York and Massachusetts in the 1970s. By the early 1980s, researchers concluded that it was a normal metabolic variant that did not warrant treatment, and eventually the NBS programs for histidinemia were closed.[54]

The advent of MS-MS in the 1990s created the possibility of including many more conditions in NBS without clear benefit to the newborn. David Millington comments that "the original paradigm for newborn screening was fairly clear cut. The disorder had to be treatable; it had to have dire consequences if not detected early. There had to be a simple and available test—the PKU paradigm we used to call it. And tandem mass spectrometry kind of interceded on top of that and created a different paradigm. It shifted into a technological paradigm, almost, and away from a considered disease-by-disease paradigm."[55] This ideological shift also brought evidential difficulties, because most diseases detectable by MS-MS are very rare. Bridget Wilcken, who pioneered the use of the technology in Australia, notes that apart from the lengthy follow-up needed to adequately estimate clinical benefit, "the rarity of most individual disorders means that it is unlikely that studies large enough to attain statistical power would ever be possible, and this is now especially important since the multiplex nature of tandem mass spectrometry has for the first time allowed extremely rare disorders to be included in a screening panel at tiny additional cost."[56]

Beginning in the 1990s, the core Wilson and Jungner tenet—that there must be an effective treatment for a disease—was being described by some as outmoded. Critics contended that treatability was too narrow a criterion. In their view, there were many other benefits for newborns, now and in the future, and also for parents. To insist that programs abide by the classical criteria is to doom us to ignorance, since the only way to develop effective treatments for rare disorders is to screen for them and learn from the experience. Moreover, infants may benefit from interventions that fall short of effective treatments. Arguing that "the old dogma cannot be allowed to stand in the way of developing effective treatments for these rare genetic disorders," Duane Alexander and Peter van Dyck, of the Health Resources and Services Administration, identified as a benefit to child and family "the potential for the child to participate in research on innovative therapies intended to prevent or to modify manifestations of the disease."[57] Another line of argument was that the family as a whole, and not only the child, should be considered a legitimate beneficiary of NBS. In this perspective, screening for conditions for which there is uncertain or even no treatment may be warranted because parents profit from prompt diagnosis for their sick child and from the opportunity to make informed reproductive choices. Moreover, it was argued, many parents consider information in itself a benefit.[58] Scott Grosse and colleagues have characterized the inclusion of conditions with much less dramatic medical benefit to the newborn than in the case of PKU, as well as the inclusion of beneficiaries other than the newborn, as a paradigm shift from a "public health emergency" to a "public health service."[59]

At the same time, those who favored a more cautious approach defended the classical criteria. They argued that the purported benefits to parents, even if real, could not ethically justify state-mandated screening. They also claimed that the enthusiasts tended to exaggerate the benefits of screening while discounting or ignoring the potential costs and harms, including unnecessary treatment of mild cases and "diseases" that might turn out to be benign, the anxieties associated with false-positive and, especially, inconclusive results generated by the increasing number of tests, and the increased demands on NBS program staff and clinicians for follow-up and management. Against the claim that screening would spare parents diagnostic odysseys, they noted that positive results may themselves "initiate diagnostic cascades, when confirmatory testing produces an ambiguous result and one test stimulates another as more questionable abnormalities are found."[60]

Such concerns are amplified as NBS proponents explore the

possibility of using genetic tests for disease susceptibility, rather than using MS-MS to identify specific metabolites that indicate disease. Whole-genome sequencing will someday be feasible for NBS and will create scenarios in which any one individual will have hundreds of "abnormal" genes that imply varying degrees of probability for future disease. Stefan Timmermans and Mara Buchbinder, who have conducted ethnographic research in a metabolic clinic as a means to explore both parents' and clinicians' experiences of expanded screening, coined the phrase "patients-in-waiting" to describe a state created when NBS identifies the possibility but does not confirm the presence of treatable disease.[61] In a sense, NBS programs have always created patients-in-waiting, as it is impossible to know, for rare conditions, the full range of clinical presentation and effect of treatment before sufficient numbers of affected people have been identified. Thus in the 1960s, clinicians struggled with whether to treat intermediate levels of hyperphenylalaninemia; each new condition added to the NBS program raised similar issues. But the number of patients-in-waiting has dramatically expanded with the advent of MS-MS and is bound to explode with technologies just over the horizon.[62]

Worldwide, the cautious approach has generally prevailed, with most jurisdictions still committed to the traditional principles for selection of new screening tests. In a thoughtful overview of NBS policy, Wilcken flatly states, "The only reason to screen is to achieve some sort of health benefit."[63] In Europe, Wilson and Jungner's criteria remain the standard for 23 of the 35 responding jurisdictions in a recent EU survey.[64] However, the criteria have clearly been interpreted quite differently in different European countries. That is perhaps not surprising, given how broadly they are formulated; judgments as to whether a condition suitable for screening is "important" and has an "acceptable" treatment, a "suitable" test, and an "adequately understood" natural history are inevitably subjective.[65] Jeffrey Botkin notes that "while the traditional Wilson and Jungner criteria were widely adopted in the field, their interpretation has been so broad as to undermine the notion that meaningful criteria exist at all."[66] Indeed, it might be said that the principles have functioned more as a resource for those who wish to constrain the expansion of screening than as a constraint per se.[67] Where there has been a will to expand, the classical screening criteria have rarely proved an insuperable barrier.

EXPLAINING THE DISPARITIES

The particularly rapid pace of expansion of newborn screening in the United States is often attributed to a "technological imperative"—the advent of MS-MS, which made it possible to screen for many conditions in a single analysis. However, though undoubtedly true that MS-MS has been a crucial factor in the drive to expand NBS, the technological imperative is, to some extent, culture-specific. After all, the same technology has been available in many countries that have only slowly expanded their screening panels, if at all. An explanation of the gap between the United States and most of the rest of the world must therefore lie elsewhere.

In a recent comparison of screening policy in the United Kingdom, the United States, and Germany, Rodney Pollitt noted that, in contrast to newborn screening policymaking in the United Kingdom, most participants in the ACMG deliberations were professionals involved in some way with the delivery of screening services, and he suggested that differences in the professional affiliations of individuals involved in NBS policymaking were probably a factor in national differences.[68] Pollitt's focus on who has a voice in making decisions can be extended beyond the ACMG recommendations to consider what forces have shaped NBS programs in the United States more generally.

One mainly (although not exclusively) American development has been the emergence of actors with a commercial stake in NBS expansion.[69] Soon after Naylor became involved in the application of MS-MS technology to newborn screening, he founded a private company that offered fee-based screening services, including tests for Duchenne muscular dystrophy and several other conditions not ordinarily provided by public health programs. NeoGen Screening, later acquired by Pediatrix Medical Group, used aggressive tactics, including patent filings, lawsuits, and legislative lobbying, to promote its expanded screening services to states, hospitals, and parents. In the late 1990s, the company (unsuccessfully) sued the New England regional screening program, claiming that it had established an illegal monopoly. In its campaign to privatize NBS, NeoGen found allies among some parents' groups and the American Legislative Exchange Council, an organization of politically conservative state legislators that adopted a "Resolution to End State-Enabled Newborn Testing Monopolies."[70] NeoGen offered to provide screening at a lower cost than public programs, and it actively lobbied legislators to require expanded screening in several states.[71] Other entities also have a financial stake in NBS. The world's largest producer of MS-MS machines is the US firm PerkinElmer.[72] That company and several nonprofit laboratories, such as the Mayo Medical Laboratories, the University of Colorado, and Baylor Medical Center, also offer fee-based NBS services. The influence of entities with a financial interest in NBS policy has been amplified through relationships with disease advocacy groups and representation on advisory bodies. For example, three of the eight members of the original SACHDNC were affiliated with entities that either sold MS-MS machines or offered fee-based newborn screening services.

At the same time that commercial interests have sought to influence policy, considerations of financial cost in health care decision making have increasingly become framed as illegitimate in the United States. Thus the Obama administration's proposal to include funding for "comparative cost-effectiveness analysis" in the 2009 economic stimulus bill was denounced by the bill's critics as "rationing" (a scare word in the United States) that "could lead to the denial of lifesaving medical treatment or euthanasia of patients."[73] According to these critics, health is "priceless" and any consideration of cost immoral, especially when children or infants are concerned. Even cost-effectiveness analysis, which compares the cost of different interventions aimed at achieving the *same* goal, is associated with socialism, totalitarianism—and Europe.[74] In response to the outcry, the phrase "cost-effectiveness analysis" was removed from the bill.[75]

Attitudes toward expansion of newborn screening reflect this larger ideological context. The EU report on current practices in NBS notes that cost considerations play a crucial role in decisions to screen or not to screen for particular conditions, and these were the leading factor in decisions not to implement screening for a particular disorder.[76] However, in the United States, cost considerations are increasingly downplayed, when not dismissed entirely. It is now common to hear that, when babies' lives are at stake, considering cost at all is morally wrong, an attitude reflected in the claim by March of Dimes officials that "a currently available test should be abandoned for a newer one, if the latter achieves a greater precision and offers a shorter turnaround time, no matter what the cost differential."[77] That is a striking reversal from the 1960s, when proponents emphasized that screening for PKU would save states money.

Critics of expansion have protested that, given limited public health budgets, resources expended on research, screening, or treatment for one disease are resources not used to cure or ameliorate the effects of other diseases or to meet non-health-related needs. In this perspective, funding decisions inevitably involve opportunity costs—interventions not undertaken that have value for others—and hence issues of distributive justice.[78] A report on expanded NBS issued by the Hastings Center typifies this perspective. Its authors use the sequel to Mississippi's expansion of newborn screening to illustrate what they see as ethically at stake when opportunity costs are ignored. In 2000, a child died from a metabolic disorder that was not then part of the state's NBS panel. After passionate lobbying by his father, the legislature passed a law that resulted in an increase from 5 to 40 in the number of included conditions. But following expansion, overall infant mortality actually increased, especially among African Americans. How could this be? The authors suggest that one reason is that, to help pay for the expansion, Mississippi doubled the fee for each newborn screened to $70. Since Medicaid covers more than half of Mississippi births, a substantial share of the resources for NBS expansion came from Medicaid funds. But the Medicaid budget was not increased; indeed, eligibility requirements were tightened, resulting in less access to prenatal and postnatal care. The authors comment that "often no one steps up to advocate for the programs that will not be undertaken and the people who will not be helped because health-care resources have been directed elsewhere . . . The advocates for [poor] women, and the children they were carrying, were either silent or ineffective."[79] But the distributive justice concerns voiced by bioethicists have had little impact on policy.

As the Mississippi case illustrates, in the United States, the action in NBS is typically at the state level, where even a few highly motivated individuals—indeed, sometimes a single individual—can have an impact on legislation, especially when opposing voices are lacking. Just as Guthrie and the NARC took advantage of the fact that state legislatures could be relatively easily moved to action by heartrending stories, especially if they involved children, so have many individuals and groups with an interest in screening expansion in more recent years. What sociologist Rachel Grob has termed the "urgency narrative"—the claim that babies' lives are at stake in decisions about whether to add new tests—has come to dominate public discourse on screening in the United States.[80] In most countries, however, a strategy that depended on powerful emotional appeals to legislators would be futile, because NBS policy is typically made at the national level—in the United Kingdom, for example, screening is embedded within the National Health Service. And even when such policy is a territorial, provincial, or regional responsibility, it is rare for the exact diseases included in NBS to be a matter of legislation.[81]

Moreover, internationally, newborn screening policy, like health policy in general, tends to be expert-driven to a much greater degree than is the case in the United States, where there is an expectation that "consumers" will be represented on advisory committees and their voices heeded. Consumer interests are often represented by patient advocacy organizations, which have come to exert an increasingly significant impact on health policy. In the domain of NBS, they have been key players—often allied with professionals who believe that "their" disease deserves much greater attention—in screening expansion. Such groups tend to be much weaker in Europe, and patients' voices in health policy more muted.[82]

Bioethicists and others who think that policy should follow the evidence have sometimes been appalled at the highly political process of medical priority setting in the United States. In their view, emotional appeals by families and disease advocacy groups have distorted the process of resource allocation, undermined structures that protect against harmful or ineffective medical interventions, and promoted unrealistic expectations of the benefits to be derived from biomedical research and applications. They worry about the representation of interests that are not easily mobilized for political action and the distributional consequences of a system that operates on the principle of "the squeaky wheel gets the grease."[83] But for better or worse (or both), that is how American politics works. The alternative is greater reliance on experts, a strategy that would be strenuously resisted and is not without costs of its own.

Yet another factor driving the expansion of NBS in the United States is the potential for lawsuits. Americans are notoriously litigious, and there are low entry barriers to the courts in the United States, where jury awards also tend to be comparatively large. As a result, concern that lawsuits might be filed by parents of a child missed by screening has always been present. Indeed, the desire to minimize such suits was a factor in the move from screening that was primarily hospital-based in the 1960s and early 1970s to screening conducted in state public health laboratories.[84] Noting that the unequal uptake of MS-MS had left the standard of care for NBS in a state of flux, geneticist-lawyer Philip Reilly remarked in 2001 that "sharp differences in standards of practice create the legal opportunity to argue that the lower standard is in fact sub-standard."[85]

THE LESSONS OF HISTORY?

Although radical changes in context have brought new players and new issues, and NBS initially involved one disease whereas current programs involve many, the controversy over expanded screening that began in the 1990s reprises to a remarkable degree that of the 1960s and 1970s. The proponents of caution still cite the scientific unknowns, the possible harms from unnecessary treatment or overtreatment, the impact of false-positive and inconclusive results, and the emphasis on tests at the expense of follow-up. They condemn the media hype and the politicized process. In general, they believe that the risks and burdens of screening are often insufficiently appreciated and the benefits consistently overstated.

The enthusiasts, on the other hand, still maintain that the natural history of many conditions will only be understood through screening; indeed, some believe that virtually all conditions detectable at birth should be screened for, since that is the most effective way to gain knowledge and develop treatments for diseases that are rare and poorly understood.[86] As in the 1960s and 1970s, enthusiasts believe that the critics' concerns are exaggerated and that it would be immoral not to spare infants and their families needless suffering. They sometimes add that even if the critics' arguments do have some merit, opposition is quixotic, since "like a snowball rolling downhill," the process "not only is unstoppable, but it grows bigger as it descends with time."[87]

In this debate, both enthusiasts and critics invoke the history of screening for PKU. Drawing lessons from that history is not new. The 1975 NAS report included an entire chapter, "Lessons Learned from the PKU Experience, and Recommendations," that interpreted the history as a cautionary tale. According to the committee, screening was mandated prematurely, as a result of overwhelming political pressure exerted by parents involved in local chapters of the NARC. In the committee's view, we should avoid repeating that kind of hurried and politicized decision making.[88] Another author similarly argued that, in the early years of screening, "thousands of infants [were] subjected to an incompletely validated and potentially hazardous intervention," but in this case we were lucky and narrowly dodged the bullet. It would be foolish indeed to make the same mistake twice.[89] That continues to be the way the early history is read by those favoring a cautious approach to expansion.

Enthusiasts generally agree with critics on the facts—they do not dispute that at the time screening was mandated, there was a host of scientific and clinical unknowns—but they interpret those facts very differently. The enthusiasts say, in effect, "Look, in the end we did fine, which shows that the worries were overblown. The problems were solved, or at least effectively managed. Fortunately, the critics in the 1960s and 70s didn't get their way. If they had, a whole generation of adults with normal intelligence would instead have suffered profound mental retardation."[90]

The lessons of history are rarely self-evident. Historical evidence, like other kinds of evidence, needs to be interpreted, and lessons and counter-lessons can be and often are derived from the same set of facts. That is hardly a unique feature of the history of newborn screening for PKU. But if accounts of this history cannot provide straightforward lessons for policy, they can expose unrecognized tensions and contradictions and enable us to understand both why they exist and why they will be difficult to resolve.

Choices related to the structure of newborn screening for PKU made in the 1960s and 1970s are embedded in our current laws, institutions, and practices and continue to provide the core framework for current newborn screening programs. This is evident in relatively mundane ways: even when there is widespread consensus on the value of expanding NBS to a new condition, for example, the decision to locate the responsibility for screening in individual states means that 50 different legislatures must amend their laws, budgets, and laboratory practices. A more profound example is the lack of requirements for informed consent for NBS. As noted in previous chapters, the issue of whether to require informed consent from parents was not debated in the 1960s, and when this issue did emerge in the 1980s, the lack of a requirement was justified on the grounds that no rational person would decline screening for a devastating disease that could be effectively treated only if diagnosed in infancy. Expansion of NBS to conditions for which diagnosis is of uncertain medical benefit to the newborn, such as fragile X syndrome or Duchenne muscular dystrophy, thus confronts a structural and normative framework shaped by assumptions specific to PKU. That framework presents similar challenges to scientists and policymakers who would like to make research use of the rich resource of dried blood spots that have generally been collected without parental consent (the subject of the epilogue). Although history cannot tell us whether we should screen for conditions such as fragile X syndrome or allow the use of residual dried blood spots for epidemiological or other research, it does help us understand why these have become such difficult and emotionally charged conundrums and the nature of the interests at stake.

EPILOGUE

"The Government Has Your Baby's DNA"

Contesting the Storage and Secondary Use of Residual Dried Blood Spots

In the half-century since the advent of the first blood-spot screening programs for phenylketonuria, fundamental changes have taken place in the context of debates about newborn screening. In the 1960s, there were few commercial interests in newborn screening, and the prevailing ethos held that it was wrong to "benefit from the public's ills." The movement for evidence-based medicine did not yet exist.[1] Bioethics, with its focus on patients' rights and the importance of informed consent, had not yet emerged as a discipline. There were no state-funded grant programs in the ethical, legal, and social implications of genetics, which in the future would focus academic and public attention on issues of informed consent, the ownership of genetic information, genetic discrimination, privacy and confidentiality, and the limitations of genetic tests. There were no multiplex technologies. No one associated newborn screening with DNA.

The many changes in political, economic, technological, and cultural context and the associated emergence of new actors have, of course, affected attitudes toward NBS. The erosion of the public health ethos that originally motivated the introduction of screening programs is particularly striking. Guthrie's politics were well to the left on the US political spectrum, and his belief in the power of government to do good was integral to his commitment to universal, compulsory screening.[2] Guthrie declined to accept royalties from the test and was always distrustful of commercial interests; thus he viewed the decision of the California NBS program to utilize existing private laboratories as an unfortunate capitulation to the state's politically powerful private pathologists.[3] Today's commonly heard claim that fee-based supplemental screening should be available to anyone willing and able to pay reflects a perspective antithetical to that which motivated Guthrie and other parent-professionals in the 1960s. Relatedly, equity arguments tend to focus on disparities in access to tests rather than access to treatment—a trend that already worried the NARC in the 1970s and has since accelerated. "Children are *suffering* and *dying* needlessly because they're born in the wrong state. A child's chances for life shouldn't be dependent on where he or she is born," argues the president of one disease advocacy group.[4] Disparities in what happens *after* a child is born rarely prompt such emotional appeals. In 1975, the NAS Committee for the Study of Inborn Errors of Metabolism asserted that "if all infants are to be screened, then there is an obligation to ensure that all infants discovered to have PKU receive optimal therapy," an assertion that has been repeated in one form or another in numerous policy documents since.[5] But there is no movement for uniform access to treatment remotely comparable to the campaign for a uniform screening panel, and universal access to treatment for NBS conditions remains highly variable among states.

Controversy has also erupted internationally about what screening programs should do with the huge number of stored dried blood-spot or "Guthrie" cards—a particularly vexing issue that no one could have anticipated before the discovery that the DNA they contained could be extracted and amplified. In 1987, Edward McCabe and colleagues found that the small amount of DNA obtained from Guthrie cards could be copied in sufficient quantity to screen for mutations in the hemoglobin gene.[6] Until then, the stored cards in the possession of screening programs had been of little interest. As an extremely large pool of DNA samples, however, those cards now acquired research, diagnostic, and forensic value. One particular appeal is that there is stored dried blood for nearly every baby in most jurisdictions, which means that researchers potentially have access to a population-based sample, rather than having to rely on volunteers who are motivated to participate in a research study. Obtaining large numbers of research participants in an unbiased manner is nearly impossible. As jurisdictions began to shift to long-term storage of dried blood spots, many more researchers sought access to the cards, a trend bolstered by the popularity of genome-wide association studies, a research approach that requires large study populations to identify genetic variations that may be associated with disease.

As with screening itself, disparities in states' storage practices widened. Today, 14 US programs, serving almost half of all infants, save the cards for more than 21 years, but some states keep them indefinitely, and some discard them almost immediately.[7] Nor is there any standard policy regarding who can use the cards and for what purposes. In the United States and elsewhere, secondary use may be limited to activities directly related to NBS program goals, principally quality control and the development of new screening tests. But the residual blood spots also have value for forensic purposes and for epidemiological and other kinds of research, and there is no consensus as to whether outsiders should be permitted access to the cards and, if so, whether they could include corporations or other for-profit entities, or whether all identifying information must be removed, thus precluding some research uses. Many jurisdictions have no explicit policies on these matters at all.[8]

Parents are rarely asked to consent to the storage and use of their babies' blood. The reason is simple. As noted in chapter 6, historically, consent has usually not been required for the screening test. At the time blood-based screening for PKU was introduced, the question of whether to require explicit consent for the test was not on the agenda. Only in the 1970s did criticism of mandatory screening become linked to the issue of informed consent. By then, the test for PKU was increasingly conceptualized as a genetic test, and in the context of the controversy over genetic testing, especially for sickle-cell disease, there was intense discussion of what principles should govern genetic screening programs. Once newborn screening was categorized as genetic screening, the lack of consent appeared incongruous, and in 1975, the NAS Committee for the Study of Inborn Errors of Metabolism recommended that participation in any neonatal genetic screening program should be at the discretion of the parents or legal guardian.

Calls for consent provisions made very little headway, however. In 1976, the state of Maryland, acting on the advice of a Commission on Hereditary Disorders, introduced a written parental consent provision. It met fierce opposition. At the time, providers voiced concern that consent requirements would undermine the program's cost-effectiveness and also fail to achieve meaningful parental choice. The state chapter of the American Academy of Pediatrics requested that the requirement be deleted out of concern that informed consent provisions would be extended to other routine procedures and that many parents would refuse permission for their infants to be tested.[9] A 1982 study found that many physicians were unaware of the provision to include parental consent; when told, 69 percent of chiefs of obstetrics and 78 percent of chiefs of pediatrics disapproved.[10]

In the 1980s and 1990s, controversy occasionally flared over the issue. But health care providers continued to argue that informed consent requirements would be costly and would undermine efficiency if the process were meaningful and would be pointless if it were not. Even prominent bioethicists, who in general favored consent requirements, argued that they were not ethically required in this case. Newborn screening, they noted, involved parental rather than personal autonomy. Parental consent is grounded in the assumption that parents know their children's interests best and are their most conscientious advocates.[11] Given the existence of an effective therapy for PKU, they concluded that the decision to reject testing poses a risk of harm to the child and can never be in his or her best interest. Thus there could be no moral basis for allowing choice. Despite the expansion of NBS to include conditions much less treatable than PKU and the emergence of justifications other than medical benefit to the child, parents are rarely offered the choice to decline screening. Almost everywhere, NBS remains de facto mandatory, even when it is not required by law. Its expansion to conditions that fall outside the PKU paradigm has generated calls for new models of informed consent for NBS programs.[12]

The residual blood spots are valuable because they contain DNA, and in many countries, there is widespread anxiety that genetic information in the wrong hands could be used for malevolent purposes. A CNN senior medical correspondent's title for a story on the lack of consent for blood-spot storage and use in the United States was "The Government Has Your Baby's DNA."[13] Commenting on a 2009 lawsuit filed against the Texas Department of Health that resulted in a settlement requiring the state to destroy the stored samples (*Beleno et al. v. Texas Department of State Health Services et al.*), a reporter writes, "The story goes beyond Andrea Beleno and the four other Texas families that filed suit against the state over the clandestine program. It gets to the heart of DNA. Who owns it? Who can access it? What does it reveal about us? How could it be used against us? What do scientists get out of our blood discards anyway?"[14]

Such concerns are hardly limited to the United States. In the Netherlands, a journalist's proposal that the screening program operated by the National Institute for Public Health and the Environment (RIVM) could use residual blood spots to identify victims of a fireworks disaster in the city of Enschede in 2000 created an uproar, with a major newspaper proclaiming, "RIVM collects DNA from Babies."[15] Although there was nothing secret about the practice of storing the blood spots, it turned out that before Enschede, almost no one who was not professionally involved in newborn screening was aware of the practice.[16] The framing of the storage issue was similar in Ireland, where, in the aftermath of a journalist's claim that the government was operating a "DNA data bank," the Office of the Data Protection Commissioner ordered the national screening laboratory to destroy 1.6 million Guthrie cards stored without parents' explicit consent.[17]

In many countries, concerns about genetic discrimination and privacy are rife; in the United States, they prompted passage of both state and federal antidiscrimination statutes. Also common are suspicions that genetic research will serve eugenic ends. On the Web and in the blogosphere, the specter of eugenics is now constantly invoked by critics of the lack of informed consent requirements for the storage and secondary use of blood spots. A story about Twila Brase, a nurse who has played a key role in protests against blood-spot storage in the United States, notes that she often refers to the film *Gattaca*, "in which genetically 'inferior' people form a social underclass, and when she testified to the state House of Representatives in 2009, she placed two books in front of her: one about the US eugenics movement, the other about the Holocaust."[18] In 2009, the Citizens' Council on Health Care, an organization Brase founded, published a report titled "Newborn Genetic Screening: The New Eugenics?" which associated NBS with compulsory sterilization laws in the United States and the Nazi campaign to rid Germany of Jews and other undesirables.[19]

In the United States and elsewhere, anxieties surrounding DNA converged with increasing suspicion of government. In the aftermath of the Enschede incident, a digital-privacy organization presented the head of the Dutch screening laboratory with its "Big Brother" award.[20] Such distrust has animated activists on both the right and left of the political spectrum. In the 2010 case of *Brearder v. State of Minnesota*, nine families alleged that the state's Genetic Privacy Act requires written parental consent for the storage and research use of NBS blood samples.[21] (Their claim was ultimately upheld by the Minnesota Supreme Court, and on January 30, 2012, the state began the process of destroying at least a million samples.)[22] The legal challenge was organized by Brase's Citizens' Council on Health Care (in 2010 renamed the Citizens' Council for Health Freedom), a libertarian group that she originally launched in opposition to the Clinton health plan. Brase wants the government out of the screening business entirely. Even after her group won its case, Brase said that she did not trust the government to follow the rules ("nobody's in there watching"). Given the risk that genetic information may not be kept private, she argued, "the best way to make sure that it never happens is to simply not get screened."[23] In her view, if parents want screening, they can always pay for it privately.

Although the basic arguments in the Texas case were similar to those in Minnesota, the suit was motivated by the opposite politics. The plaintiff was the Texas Civil Rights Project, whose mission is "to promote racial, social, and economic justice through litigation, education, and social services for low/moderate-income persons least able to defend themselves."[24] Its heroes include Cesar Chavez and Noam Chomsky, and the targets of its lawsuits are typically local and state police and the Border Patrol. Although the Texas Civil Rights Project stands at the opposite end of the political spectrum from the Citizens' Council for Health Freedom, the organizations share a distrust of the state and especially the fear that it will misuse any DNA it can manage to obtain. As in Minnesota, the plaintiffs prevailed; in this case, the Texas health department agreed to destroy all five million residual blood spots for which consent had not been obtained.[25]

Of course, there has been pushback. Professional societies, advocacy groups, corporations, public health officials, and others involved in NBS have rallied to defend the storage and research use of residual blood spots. They invariably stress that "newborn screening saves lives."[26] The sole goal of NBS, as described in this literature, is to protect babies at high risk of "developing diseases with high morbidity and mortality."[27] In *this* campaign, there is no mention of untreatable conditions or of aims other than medical benefit to the infant. According to the Preserving the Future of Newborn Screening Coalition, "many babies detected through newborn screening seem healthy at birth but in fact have diseases that will cause them to become very sick in infancy, or even later in childhood, without treatment. The diseases that can be detected through newborn screening can cause death or permanent disability if not treated promptly. Newborn screening allows these babies to be identified shortly after birth and get the treatment they need so they can live happy, healthy, productive lives."[28] This is the newborn screening envisaged by Wilson and Jungner, where the ability to treat is fundamental. It is the PKU paradigm, invoked when useful.

ACKNOWLEDGMENTS

Nearly two decades have elapsed since Diane Paul began to research the history of PKU. Over such an extended period, one amasses many debts, some of which she might no longer even recall. To anyone inadvertently excluded from the list of those to whom thanks are due, she offers sincere apologies.

Of the debts that she knows she owes, the first and foremost is to Paul J. Edelson, MD, a pediatrician with a long-standing interest in the history of PKU. Paul and Diane had coauthored an essay and also planned to collaborate on this book. However, while the research was still in its early stages, Paul accepted a position with the Centers for Disease Control, where his schedule made it necessary for him to withdraw from the project. Generous as always, he allowed Diane to make use of text that reflected his contributions to their joint work and provided unwavering support and encouragement from the sidelines.

The book owes its existence to an intervention by French historian of medicine Ilana Löwy. After many years of research, Diane found herself overwhelmed by the amount and variety of material she had amassed and was hopelessly lost in conceptual thickets. Ilana suggested a way to make the story coherent. At an earlier stage, Ilana also had insisted that the research include patient interviews. Although Diane cursed her many times when wrestling with the need to obtain approvals and yearly reapprovals from both university and hospital Institutional Review Boards, she came to believe that the effort was worth it and that a history of PKU without patients' voices would be seriously incomplete.

Special thanks are also due to the staff of the Program in Metabolism at Boston Children's Hospital. Diane became involved with the group almost at the start of her research when she arranged to talk with psychologist Susan E. Waisbren, who had done seminal work on psychosocial issues in PKU. Even though the discussion ended in an argument, Susan—perhaps hopeful that a closer look at newborn screening would correct at least some of Diane's misperceptions—suggested that Diane attend one of the monthly meetings of the clinic's planning group. She did so and continued, over the next eighteen years, to participate in these meetings and related activities. What she learned from this association was invaluable, as were the aid and encouragement provided by program staff, especially Susan, nutritionist Fran Rohr, and pediatrician and biochemical geneticist Harvey L. Levy. Susan, Fran, and Harvey invariably responded readily to what must have seemed a never-ending stream of queries about newborn screening past and present, as well as requests for comments on several draft chapters. They were doubtless as glad as Diane when the book was finally finished.

Others who provided helpful remarks on individual chapters include Mary Ampola, Donna Messner, Donna Rhodes, and on aspects of newborn screening in Europe, Martina Cornel, Carla van El, and Gerard Loeber. Special thanks are owed to Neil R. M. Buist, pediatrician, biochemical geneticist, and former director of the Pediatric Metabolic Laboratory at Oregon Health Services University. Illustrating the old adage that no good deed goes unpunished, he provided particularly extensive and useful commentary on one chapter and was thus prevailed upon to read several others—in multiple iterations.

Several archivists and librarians went well beyond the call of duty in tracking down documents and photographs. Diane is especially grateful to Dorothy Barr, Public Services Librarian, Ernst Mayr Library of the Museum of Comparative Zoology, Harvard University; Judy Engelberg, Archivist, Joseph P. Kennedy, Jr. Foundation; William Offhaus, Senior Staff Assistant, University Archives, University at Buffalo, State University of New York; Olivia Pickett, Director of Library Services, MCH Library, National Center for Education in Maternal and Child Health, Georgetown University; Donna Carcaci Rhodes, Curator, Pearl S.

Buck National Historic Landmark Home; and David Rose, Archivist, March of Dimes.

At Harvard's Museum of Comparative Zoology, thanks are due to Richard Lewontin for help in explaining genetic theories and concepts and to Jenn Thompson for assistance with all matters computer-related. Diane also thanks the Guthrie family for their encouragement and assistance in obtaining photographs.

Support for Diane's research was provided by both the National Science Foundation, Grant No. SBR-9511909 (1996–97), and the National Institutes of Health, Grant No. 1R03HG003730-01 (2005–7). She also benefited from a fellowship from the CSG Centre for Society and the Life Sciences, which allowed her to spend the winter of 2012 as a visitor in the Community Genetics section at the VU University Medical Center in Amsterdam. Martina Cornel and members of her research team, including Carla van El, Pascal Borry, Lidewij Henneman, Stephanie Weinreich, Sarah van Teeffelen, and Tessel Rigter, did their best to educate Diane about newborn screening policies and perspectives in Europe. Thanks for other crucial types of support are due to Maud Radstake, CSG Research and Dialogue Manager; Maria-Lucia Cantore, CSG Office Manager; and Wilma Ijzerman-Lap, Secretary for Community Genetics.

Both professionals and individuals with PKU were interviewed for this book. Diane is particularly grateful to those current and former patients of the metabolic clinic at Boston Children's Hospital who agreed to share their experiences with a stranger. She is also indebted to the many metabolic researchers and clinicians and the newborn-screening policymakers who agreed to be interviewed and, in some cases, re-interviewed. Because this is a short book intended for a nonspecialist audience, it was necessary to summarize and simplify much of what was learned from specialists. Diane apologizes for any errors and hopes that these experts, who gave so generously of their time, knowledge, and insights, understand that interviews provided important background information even when not directly cited. She also thanks sociologist Robert S. Weiss, author of the superb *Learning from Strangers*, whose coaching helped her become both a more perceptive questioner and a more attentive listener. (She had a long way to go!)

Given that Diane has been writing about the history of PKU for almost two decades, some of the ideas in this book inevitably appeared in earlier publications. These include an essay coauthored with Paul Edelson, "The Struggle over Metabolic Screening," in Soraya de Chadarevian and Harmke Kamminga, eds., *Molecularising Biology and Medicine: New Practices and Alliances, 1930s–1970s* (Reading: Harwood Academic Publishers, 1997), 203–20; "The History of Newborn Phenylketonuria Screening in the U.S.," in Neil A. Holtzman and Michael S. Watson, eds., *Promoting Safe and Effective Genetic Testing in the United States* (Baltimore: Johns Hopkins University Press, 1998), 137–60; "PKU Screening: Competing Agendas, Converging Stories," in Michael Fortun and Everett Mendelsohn, eds., *The Practices of Human Genetics* (Dordrecht: Kluwer, 1999), 185–96; and "Contesting Consent: The Challenge to Compulsory Neonatal Screening for PKU," *Perspectives in Biology and Medicine* 42 (1999): 207–19. Diane also gratefully acknowledges permission from Springer to excerpt text that originally appeared in "From Reproductive Responsibility to Reproductive Autonomy," in Lisa S. Parker and Rachel A. Ankeny, eds., *Mutating Concepts, Evolving Disciplines: Genetics, Medicine, and Society* (Dordrecht: Kluwer, 2002), 87–105; and the journal *Medicina nei Secoli* for permission to use material from "Contested Conceptions: PKU in the Postwar Discourse on Reproduction," *Medicina nei Secoli* 14, no. 3 (2002): 773–91. Chapter 6 has been substantially adapted for publication as "How PKU Became a Genetic Disease," in Bernd Gausemeier, Staffan Müller-Wille, and Edmund Ramsden, eds., *The History of Human Heredity in the Twentieth Century* (London: Pickering and Chatto, 2013).

Jeff Brosco's first—and really, only—acknowledgment goes to his coauthor, Diane, for whom this publication reflects decades of research. Jeff is forever grateful to Diane for her generosity in sharing her years of labor. He would also like to thank the key teachers whose mentorship helped develop his dual interests in history and developmental pediatrics: Robert Kohler, Henrika Kuklick, Charles Rosenberg, Rosemary Stevens, R. Rodney Howell, and F. Daniel Armstrong. For portions of this work, he was supported by grants from the Robert Wood Johnson Foundation General Scholar Program, Grant No. 033954 (1998–2003); the Maternal Child Health Bureau Advisory Committee on Heritable Disorders and Genetic Diseases in Newborns and Children (Department of Health and Human Services, 2006–7) Subcontract No. 06-C210-01; and an Arsht Distinguished Ethics Faculty Award at the University of Miami (2012–14).

Finally, Diane and Jeff jointly thank series editor Charles Rosenberg for his faith in the project, an anonymous reviewer for insightful comments, and the staff at Johns Hopkins University Press for their wise counsel, unflagging patience, and consistent good humor. It was a particular delight to work with executive editor Jacqueline Wehmueller, editorial assistant Sara Cleary, copyeditor Linda Strange, and production editor Courtney Bond.

A NOTE ON SOURCES

Patient Interviews

No "biography" of PKU, even a short one, would be complete if it did not pay attention to the lived experience of patients and their families. Indeed, one aim of this book is to provide a more realistic account of life with the disease than readers are likely to encounter in discussions intended for nonspecialists. To that end, between January 2006 and June 2008, Diane Paul conducted in-depth interviews with eighteen adolescents and adults recruited through the metabolic clinic at Boston Children's Hospital.

In identifying prospective participants, an effort was made to maximize the range of age, health, marital and socioeconomic status, and education. Those especially well able to describe their beliefs and experiences were oversampled. Since the problem of maternal PKU was of particular interest, thirteen of the interviewees were female. The aim of this study was not to generalize—which would have been impossible, given the small numbers and geographically and otherwise unrepresentative sample—but rather to obtain insights into the meaning of PKU from people who live with it (understanding that we are describing the experiences, thoughts, and feelings of people in this particular sample) and also to include patients' voices in our narrative, which we believed would increase its richness and complexity.

After Institutional Review Board approvals were obtained from both Boston Children's Hospital and the University of Massachusetts Boston (UMB), Susan E. Waisbren, the senior psychologist for the metabolism program and principal investigator for this component of the larger study, reviewed prospective participants with other clinic staff and with Diane. Clinic staff then made an initial inquiry as to whether these individuals might be interested in participating in the study. Positive responses were followed up with a letter cosigned by Diane and Susan, which described the aims of the study in more detail. Aliases are used for all participants referred to by name in the book Apart from gender and approximate age at the time of the interview, all personal identifiers have been omitted.

Insights derived from these formal interviews have been supplemented with material from newsletters, websites, and a variety of other publicly available sources. Over the past eighteen years, Diane has also attended the monthly planning-group meeting of the New England Consortium of Metabolic Programs and engaged in various outreach activities, such as helping to staff an annual PKU summer camp, which provided a more informal entrée into the real world of PKU. Diane also ran a StoryCorps-type booth, sponsored by the organization Patient Power, at the first International Conference for Teens and Adults with PKU, held in Chicago, August 15–16, 2008. The stories were recorded, with the aim of posting them on the Patient Power website, and CDs were distributed to conference participants. Although the stories were of a semi-public character, they were ultimately not published on the website, and Diane decided to anonymize any individuals quoted in this text, who are described only in very general terms like "a young man."

Oral History Interviews

Between 2006 and 2009, Diane also conducted a series of oral history interviews with individuals (listed below) who have long experience with one or more aspects of newborn screening, including its policy, laboratory, research, and clinical dimensions. Permission to conduct these interviews was granted by the UMB Institutional Review Board. Nearly all participants consented to the deposit of the recordings and transcripts of their interviews at the American Philosophical Society in Philadelphia, where they will be included in its Genetics Project and made available to other scholars.

Phyllis B. Acosta, Atlanta, GA, December 7, 2007
Duane Alexander, Bethesda, MD, March 14, 2006
Samuel P. Bessman, Los Angeles, CA, March 22, 2006
Neil R. M. Buist, Portland, OR, September 4, 2007
Stephen D. Cederbaum, Los Angeles, CA, May 7, 2009
George C. Cunningham, Berkeley, CA, March 30, 2007
Robert O. Fisch, Minneapolis, MN, March 9, 2007
William B. Hanley, Toronto, ON, December 11, 2006
W. Harry Hannon, Atlanta, GA, November 17, 2008
Neil A. Holtzman, Baltimore, MD, December 13, 2006
Ernest B. Hook, San Raphael, CA, April 1, 2007
Kathleen Huntington, Portland, OR, September 5, 2007
Henry N. Kirkman, Chapel Hill, NC, February 12, 2006
Richard Koch and Jean H. Koch, Los Angeles, CA, March 25, 2006
Harvey L. Levy, Boston, MA, December 8 and 21, 2006
Edward R. B. McCabe and Linda McCabe, Los Angeles, CA, February 22, 2007
David Millington, Raleigh-Durham, NC, December 17, 2008
William H. Murphey, Portland, OR, September 4, 2007
Edwin W. Naylor, Isle of Pines, SC, October 23, 2008
Frances J. Rohr, Boston, MA, March 7, 2007
Charles R. Scriver, Montreal, QC, November 13, 2008
Rani H. Singh, Atlanta, GA, December 7, 2007
Bradford L. Therrell, Austin, TX, November 24, 2008
Christine N. Trahms, Seattle, WA, September 7, 2007

A Note on Sources

Judith M. Tuerck, Portland, OR, September 5, 2007
Susan E. Waisbren, Boston, MA, February 8, 2007
Louis I. Woolf, Vancouver, BC, June 15, 2009

More informal, background interviews were also conducted in person or by phone with Jessie Davis (New York, NY), Felix de la Cruz (Bethesda, MD), Gerard Loeber (Amsterdam, Netherlands), Judith Swazey (Bar Harbor, ME), Hilary Valance (Vancouver, BC), and Francjan van Spronsen (Groningen, Netherlands). Jeffrey Brosco also conducted a short telephone interview with Robert E. Cooke (Vero Beach, FL).

Manuscript Collections and Personal Papers

The following collections were especially important:

Records of the US Children's Bureau (USCB), US Department of Health, Education, and Welfare. Most of the very extensive USCB materials related to the field trials of the Guthrie test, the establishment of newborn screening in the United States and abroad, and the PKU Collaborative Study are stored at National Archives and Records Administration II (NARA II) in College Park, Maryland. However, about 450 USCB documents, including several dozen related to PKU, were transferred to the National Center for Education in Maternal and Child Health (NCEMCH), Georgetown University, Washington, DC. The published documents were recently digitized and are available at www.mchlibrary.info/history/childrensbureau.html. Other archived materials remain in storage and have only brief inventories. All USCB items in both collections are from Record Group 102.

Robert Guthrie Phenylketonuria (PKU) Papers, University Archives, State University of New York at Buffalo. This collection consists of thirty boxes of largely unsorted materials.

Elizabeth Boggs Papers, Elizabeth M. Boggs Center on Developmental Disabilities, University of Medicine and Dentistry of New Jersey–Robert Wood Johnson Medical School, New Brunswick, NJ. (This collection was previously housed at the Walter E. Fernald Developmental Center's Samuel Gridley Howe Library in Waltham, MA.)

Archives of the Joseph P. Kennedy, Jr., Foundation, Washington, DC

Archives of the March of Dimes Foundation, White Plains, NY

Lionel S. Penrose Papers, University College London, London

Other collections consulted include the papers of Gunnar Dybwad, now in the Robert D. Farber University Archives and Special Collections Department, Brandeis University, Waltham, MA; the Ava Helen and Linus Pauling Papers, Special Collections, Oregon State University, Corvallis, OR; the Asbjørn Ivar Følling correspondence in the Rockefeller Foundation Archives, Rockefeller Archive Center, Sleepy Hollow, NY; Elizabeth Boggs Personal Papers (relating to the work of the President's Panel on Mental Retardation) at the John F. Kennedy Presidential

Library and Museum, Boston, MA; and the personal papers of science journalist Shirley Sirota Rosenberg, Washington, DC, of metabolic clinician-researchers Harvey L. Levy, Boston, MA, and Neil R. M. Buist, Portland, OR, and of screening laboratory director George Murphey, Portland, OR.

NOTES

Frequently cited collections are identified by the following abbreviations:

Guthrie papers	Robert Guthrie Phenylketonuria (PKU) Papers, University Archives, State University of New York at Buffalo, Buffalo, NY
USCB, NARA II	Records of the US Children's Bureau, Record Group 102, US Department of Health, Education, and Welfare, National Archives and Records Administration, College Park, MD
USCB, NCEMCH	Records of the US Children's Bureau, Record Group 102, US Department of Health, Education, and Welfare, National Center for Education in Maternal and Child Health, Georgetown University, Washington, DC

Preface

1. Incidence varies with region and ethnicity, ranging from 1 in 2,600 births in Turkey to 1 in 125,000 in Japan. Estimates of incidence (and prevalence) are very rough, given the variation among jurisdictions in defining PKU. National Institutes of Health (NIH), "Phenylketonuria: Screening and Management," Consensus Development Conference Statement (Oct. 16–18, 2000), consensus.nih.gov/2000/200 0phenylketonuria113html.htm, 1–27; statistic for the United States on p. 1.

2. Jeffrey P. Brosco, Michael Mattingly, and Lee M. Sanders, "Impact of Specific Medical Interventions on Reducing the Prevalence of Mental Retardation," *Archives of Pediatric and Adolescent Medicine* 160 (2006): 302–9.

3. Angus J. Clarke, "Newborn Screening," in *Genetics, Society, and Clinical Practice*, ed. Peter S. Harper and Angus J. Clarke (Oxford: Bios Scientific Publishers, 1997), 107–17, on 108.

4. H. Christina Fan, Wei Gu, Jianbin Wang, et al., "Non-invasive Prenatal Measurement of the Fetal Genome," *Nature* 487 (2012): 320–24, on 324. Diane Paul thanks Robert Resta for bringing this article to her attention.

5. Douglas Dales, "Infant PKU Tests Made Mandatory," *New York Times*, Apr. 25, 1964, 31–32, on 32.

6. Chris Feudtner, *Bittersweet: Diabetes, Insulin, and the Transformation of Illness* (Chapel Hill: University of North Carolina Press, 2003).

7. We refer readers who would like to know more about the biochemistry and

genetics of PKU, including contested theories of its pathogenesis, to the fine book by biochemist Seymour Kaufman (1924–2009), a former chief of the laboratory of neurochemistry at the National Institute of Mental Health, whose extensive research on PKU included confirming that a deficiency in the phenylalanine hydroxylase (PAH) enzyme was responsible for the condition. Seymour Kaufman, *Overcoming a Bad Gene: The Story of the Discovery and Successful Treatment of Phenylketonuria, a Genetic Disease That Causes Mental Retardation* (AuthorHouse, 2004). New and emerging treatments are discussed by Kaufman in greater detail in a recent review essay by Susan E. Waisbren, "Phenylketonuria," in *Handbook of Neurodevelopmental and Genetic Disorders in Children*, 2nd ed., ed. Sam Goldstein and Cecil R. Reynolds (New York: Guildford Press, 2011), 398–424; and in Amaya Bélanger-Quintana, Alberto Burlina, Cary O. Harding, and Ania C. Muntau, "Up to Date Knowledge on Different Treatment Strategies for Phenylketonuria," *Molecular Genetics and Metabolism* 104, suppl. (2011): S19–25.

8. Wylie Burke, Beth Tarini, Nancy A. Press, and James P. Evans, "Genetic Screening," *Epidemiologic Reviews* 33, no. 1 (2011): 148–64, on 149. The authors reference Daniel J. Kevles, *In the Name of Eugenics: Eugenics and the Uses of Human Heredity* (Cambridge, MA: Harvard University Press, 1995). A different perspective informs Jean-Paul Gaudillière, "Reframing Pathological Heredity: Pedigrees, Molecules, and Genetic Counseling in Postwar France," *Alter: European Journal of Disability Research* 5, no. 1 (2011): 7–15.

9. Jeffrey P. Brosco, "More Than the Names Have Changed: Exploring the Historical Epidemiology of Intellectual Disability in the US," in *Healing the World's Children: Comparative and Interdisciplinary Approaches to Child Health in the Twentieth Century*, ed. Cynthia Comacchio, Janet Golden, and George Weisz (Montreal: McGill-Queens University Press, 2008), 205–34.

10. Robert L. Schalock, Sharon A. Borthwick-Duffy, Wil H. E. Buntinx, et al., *Intellectual Disability: Definition, Classification, and Systems of Supports*, 11th ed. (Washington, DC: American Association on Intellectual and Developmental Disabilities, 2010).

11. World Health Organization (WHO), *International Classification of Functioning, Disability, and Health* (Geneva: WHO, 2001).

12. Kansas University Center for the Study of Family, Neighborhood and Community Policy, *Usage of the Term "Mental Retardation": Language, Image, and Public Education* (Lawrence: Kansas University Center for the Study of Family, Neighborhood and Community Policy, 2002).

13. Brosco, "More Than the Names Have Changed."

Introduction. Pearl Buck, PKU, and Mental Retardation

1. Pearl S. Buck, "The Child Who Never Grew," *Ladies Home Journal* 67 (May 1950): 35, 146–60, 152, 154, 156, 159–60, 163–65, 167, 169; Pearl S. Buck, *The Child Who Never Grew*, 2nd ed., foreword by James A. Michener; introduction by Martha M.

Jablow; afterword by Janice C. Walsh (Bethesda, MD: Woodbine House, 1992 [New York: John Day, 1950]).

2. Buck, *Child Who Never Grew*, 41.

3. On the Buck family's sometimes harrowing experiences in China, see Hilary Spurling, *Pearl Buck in China: Journey to The Good Earth* (New York: Simon and Schuster, 2010), esp. 152–61.

4. Donna Carcaci Rhodes, Curator, Pearl S. Buck National Historic Landmark Home, notes that John Lossing Buck was very much a part of this process but that Pearl Buck, who was divorced from "Lossing" at the time she wrote the book, erased him from both her life and her writing. Personal communication, Dec. 6, 2012.

5. Peter Conn, *Pearl S. Buck: A Cultural Biography* (Cambridge: Cambridge University Press, 1996), 111; James W. Trent, Jr., *Inventing the Feeble Mind: A History of Mental Retardation in the United States* (Berkeley: University of California Press, 1994), 231.

6. Stanley Finger and Shawn E. Christ, "Pearl S. Buck and Phenylketonuria (PKU)," *Journal of the History of the Neurosciences* 13, no. 1 (2004): 44–57.

7. Janice C. Walsh, "Afterword," in Buck, *Child Who Never Grew*, 91.

8. The exact date of the diagnosis is not known, but Willard Centerwall, who in the early 1950s developed the "wet-diaper" test for PKU, reported that when he visited Buck at her home in Pennsylvania in 1960, she mentioned that Carol had recently been diagnosed as a result of testing at Vineland. Siegried A. Centerwall and Willard R. Centerwall, "The Discovery of Phenylketonuria: The Story of a Young Couple, Two Retarded Children, and a Scientist," *Pediatrics* 105, no. 1 (2000): 89–103, on 89.

9. That there is nothing simple about "simple" monogenic disorders like PKU is argued by Charles R. Scriver and Paula J. Waters in "Monogenic Traits Are Not Simple: Lessons from PKU," *Trends in Genetics* 15 (1999): 267–72. On numbers of types of mutations at the PAH locus, see Manyphong Phommarnih and Charles Scriver, "Phenylalanine Hydroxylase Mutation Map" (rev. Jan. 8, 2007), PAH db Phenylalanine Hydroxylase Locus Knowledgebase, www.pahdb.mcgill.ca/Information/MutationMap/mutationmap.pdf.

10. The mild form of the disease, in which there is residual enzyme activity and hence only slightly elevated levels of blood phenylalanine, is called hyperphenylalaninemia, as distinguished from "classical PKU." Whether hyperphenylalaninemia should be considered benign and thus not treated has historically been a matter of controversy.

11. Trent, *Inventing the Feeble Mind*, 229–31.

12. Buck, *Child Who Never Grew*, 35.

13. Kathleen W. Jones, "Education for Children with Mental Retardation: Parent Activism, Public Policy, and Family Ideology in the 1950s," in *Mental Retardation in America: A Historical Reader*, ed. Steven Noll and James W. Trent (New York: New York University Press, 2004).

14. "Education: Lift Up Your Head . . . ," *Time* 56 (July 24, 1950): 41–42; *Reader's Digest* 57 (Sept. 1950): 18–25; Martha M. Jablow, "Introduction," in Buck, *Child Who Never Grew*, 3.

15. Buck, *Child Who Never Grew*, 27.

16. Buck wrote, "In Carol's case nothing matters, it is too late. But I think of your children, who carry the genes in their bodies. It is essential before they marry, that this blood is tested, and the blood of the person they marry. The test is relatively simple. If their marriage carries any phenylketonuria, however, one in every four children, at least, of those known to them, will be like Carol. I was in Norway in September and met Dr. Følling, who has done most of the research on phenylketonuria. I felt I should share the information with you, for the sake of your two children." (Lossing's children would have had a 50% chance of being carriers.) Excerpts from this 1959 letter were generously provided by Donna Carcaci Rhodes.

17. Walsh, "Afterword," 97.

18. Buck, *Child Who Never Grew*, 87.

19. Charles R. Scriver, "The PAH Gene, Phenylketonuria, and a Paradigm Shift," *Human Mutation* 28, no. 9 (2007): 831–45, on 832.

20. Charles R. Scriver, interview, Montreal, Nov. 13, 2008. (All interviews were with DBP.)

21. Henry N. Kirkman, "Projections of a Rebound in Frequency of Mental Retardation from Phenylketonuria," *Applied Research in Mental Retardation* 3 (1982): 319–28, on 326.

Chapter 1. The Discovery of PKU as a Metabolic Disorder

1. Asbjørn Følling, "The Original Detection of Phenylketonuria," in *Phenylketonuria and Some Other Inborn Errors of Amino Acid Metabolism*, ed. Horst Bickel, Fred P. Hudson, and Louis I. Woolf (Stuttgart: Georg Thieme, 1971), 1–3, on 1; Siegried A. Centerwall and Willard R. Centerwall, "The Discovery of Phenylketonuria: The Story of a Young Couple, Two Retarded Children, and a Scientist," *Pediatrics* 105, no. 1 (2000): 89–103, on 95.

2. The fellowship experience resulted in the offer of a position at Harvard's Fatigue Laboratory, which had recently been co-founded by Henderson and David Bruce Dill. Følling declined this when the Norwegians counter-offered, in 1930, with a professorship in pathological physiology at the University of Oslo, a position especially created for him with parliamentary approval. "Grant Action 37076," folder 15, box 1, series 767, record group 1.1, and "Div Corr Følling," folder 69, box 6, series 1.1, record group 6.1, both in Rockefeller Foundation Archives, Rockefeller Archive Center, Sleepy Hollow, NY. See also Centerwall and Centerwall, "Discovery of Phenylketonuria," 94; Shawn E. Christ, "Asbjørn Følling and the Discovery of Phenylketonuria," *Journal of the History of the Neurosciences* 12 (2003): 44–54, on 46, 48. Diane Paul is grateful to Tore Tennøe for alerting her to the existence of material on Følling at the Rockefeller Archive Center.

3. Archibald E. Garrod, *Inborn Errors of Metabolism* (1909), reprinted in *Garrod's Inborn Errors of Metabolism*, with a supplement by H. Harris (London: Oxford University Press, 1963), 13.

4. This interpretation was probably first suggested by William Bateson, with whom Garrod had corresponded. Alexander G. Bearn, *Archibald Garrod and the Individuality of Man* (Oxford: Oxford University Press, 1993), 53–62.

5. Archibald E. Garrod, Charles R. Scriver, and Barton Childs, *Garrod's Inborn Factors in Disease*, including an annotated facsimile reprint of Archibald E. Garrod's *Inborn Factors in Disease* (London: Oxford University Press, 1989), 123.

6. Donna A. Messner notes that Følling had already tested for diabetes and that the ferric chloride test had never been associated with mental retardation. So why did he use it? She speculates that his dual training in medicine and chemistry would lead him, in thinking about the children's odor, to want to test for conditions associated with substances that could produce distinctive smells and that he knew diabetic ketosis is associated with a particular scent. Donna A. Messner, "On the Scent: The Discovery of PKU," *Chemical Heritage*, spring 2012, 32–36, on 34.

7. Phenylpyruvic acid is odorless; the characteristic smell of the urine was due to the presence of a close chemical relative, phenylacetic acid. Seymour Kaufman, *Overcoming a Bad Gene: The Story of the Discovery and Treatment of Phenylketonuria, a Genetic Disease That Causes Mental Retardation* (AuthorHouse, 2004), 7.

8. Asbjørn Følling, "Über Ausscheidung von Phenylbrenztraubensäure (ppa) in den Harn als Stoffwechselanomalie in Verbindung mit Imbezillität," *Hoppe-Seyler's Zeitschrift für Physiologische Chemie* 227 (1934): 169–76; reprinted in Samuel Boyer, IV, ed., *Papers on Human Genetics* (Englewood Cliffs, NJ: Prentice-Hall, 1963), 95–102.

9. Kaufman, *Overcoming a Bad Gene*, 8.

10. K. Floss and Asbjørn Følling, "Über das Verhalten der Phenylmilchsäure in vitro und in vivo," *Hoppe-Seyler's Zeitschrift für Physiologische Chemie* 254 (1938): 250–55; Asbjørn Følling and K. Closs, "Über das Vorkommen von l-Phenylalanin in Han und Blut bei Imbecillitas Phenylpyrouvica," *Hoppe-Seyler's Zeitschrift für Physiologische Chemie* 254 (1938): 115–16; Følling, "Original Detection"; Christ, "Asbjørn Følling," 48. The effect was confirmed by Lionel Penrose and Juda Hirsch Quastel in "Metabolic Studies in Phenylketonuria," *Biochemical Journal* 31 (1937): 266–74.

11. Phenylpyruvic acid is present in small quantities in the urine of many people. Neil R. M. Buist, personal communication, Nov. 22, 2011.

12. New England Consortium of Metabolic Programs, Video Library, "Discovery of Phenylketonuria (PKU) by Dr. Asbjørn Følling," newenglandconsortium.org/for-professionals/video-library.

13. Asbjørn Følling, O. Lous Mohr, and Lars Ruud, "Oligophrenia Phenylpyrouvica: A Recessive Syndrome in Man," *Skrifter Det Norske Vitenskapsakademi I Oslo. I. Mat Naturv Klasse* 13 (1945): 1–44; Følling, "Original Detection," 2; Ragnar Følling Elgjo, "Asbjørn Følling, His Life and Work," in *Medical Genetics: Past, Present, Future*, ed. Kåre Berg (New York: Alan R. Liss, 1985), 79–89, on 85.

14. Lionel S. Penrose, *Biology of Mental Defect*, rev. ed., preface by J. B. S. Haldane (London: Sidgwick and Jackson, 1954), 143; Christ, "Asbjørn Følling," 49.

15. One was a patient and the other was his non-institutionalized but impaired brother. Lionel S. Penrose, "Two Cases of Phenylpyruvic Amentia," *Lancet* 225 (1935): 23–24.

16. Lionel S. Penrose, "Inheritance of Phenylpyruvic Amentia (Phenylketonuria)," *Lancet* 226 (1935): 192–94.

17. Penrose, *Biology of Mental Defect*, 143; Lionel S. Penrose, *Outline of Human Genetics* (New York: John Wiley and Sons, 1959), 40.

18. Lionel S. Penrose, "Memoirs—1964: Phenylketonuria" (unpublished ms.), Penrose Papers, University College London, London, file 72/2, p. 4.

19. See George A. Jervis's "The Genetics of Phenylpyruvic Oligophrenia," *Journal of Mental Science* 85 (1939): 719–62; "Studies on Phenylpyruvic Oligophrenia," *Journal of Biological Chemistry* 169 (1947): 651–56; and "Phenylpyruvic Oligophrenia Deficiency of Phenylalanine Hydroxylating System," *Proceedings of the Society for Experimental Biology and Medicine* 82 (1953): 514–15; George A. Jervis, Richard J. Block, Diana Bolling, and Edna Kanze, "Chemical and Metabolic Studies on Phenylalanine: II. The Phenylalanine Content of the Blood and Spinal Fluid in Phenylpyruvic Oligophrenia," *Journal of Biological Chemistry* 134 (1940): 105–13.

20. This point about the possible benefits of dietary phenylalanine exclusion was not universally appreciated in the early years of screening. Horst Bickel and Werner Grueter, "The Dietary Treatment of Phenylketonuria—Experiences during the Past 9 Years," in *Mental Retardation: Proceedings of the First International Medical Conference at Portland, ME*, ed. Peter W. Bowman and Hans V. Mautner (New York: Grune and Stratton, 1960), 272.

21. Even today, there is no consensus as to whether the damage in PKU is caused directly by phenylalanine's toxic effects on the brain, or indirectly by one of its abnormal metabolites, or even more indirectly by its interference with the ability of other amino acids to enter the brain. For an excellent discussion of the current state of play, see Kaufman, *Overcoming a Bad Gene*, 158–72.

22. Penrose, "Memoirs," 5; Harry Harris, "Lionel Sharples Penrose, 1898–1972," *Biographical Memoirs of Fellows of the Royal Society* 19 (1973): 521–61, on 529; Daniel J. Kevles, *In the Name of Eugenics: Genetics and the Uses of Human Heredity* (Cambridge, MA: Harvard University Press, 1995), 177–78; Renata Laxova, "Lionel Sharples Penrose, 1898–1972: A Personal Memoir in Celebration of the Centenary of His Birth," *Genetics* 150, no. 4 (1998): 1333–40. In a letter to Hopkins of January 26, 1935, Penrose also noted that Quastel was trying to devise a quantitative test for phenylalanine so that they could "experiment with various kinds of diet." Penrose Papers, University College London, London, file 139/8.

23. A measured amount of the phenylalanine could then be added back to the diet. As reported by William L. Laurence in "Find New Chemical to Sift Life Secrets," *New York Times*, Apr. 4, 1939, 21, "Attempts will be made . . . to formulate a

diet containing a very small amount of phenylalanine. On such a diet it is hoped that the brains of patients with this hereditary defect may develop normally." See also "Hope of Curing Tuberculosis, Influenza and Leprosy," *Science News-Letter* 35, no. 15 (1939), 234–36, on 235.

24. Ruth Brecher and Edward Brecher, "Saving Children from Mental Retardation," *Saturday Evening Post*, Nov. 21, 1959, 32–33, 109–11, on 109.

25. Følling, "Original Detection," 2.

26. Jervis, "Genetics," 760.

27. Penrose, "Memoir," 7. As it turned out, however, everyone excretes a small amount of phenylpyruvic acid.

28. The literature on Penrose and eugenics is extensive. A sampling: Harris, "Lionel Sharples Penrose"; Pauline M. H. Mazumdar, *Eugenics, Human Genetics, and Human Failings: The Eugenics Society, Its Sources, and Its Critics in Britain* (London: Routledge, 1992), esp. chap. 4; David C. Watt, "Lionel Penrose, F.R.S. (1898–1972) and Eugenics: Part One," *Notes and Records of the Royal Society London* 52, no. 1 (1998): 137–51; David C. Watt, "Lionel Penrose, F.R.S. (1898–1972) and Eugenics: Part Two," *Notes and Records of the Royal Society London* 52, no. 2 (1998): 339–54; Daniel J. Kevles, "Lionel Penrose, Mental Deficiency and Human Genetics," in *Fifty Years of Human Genetics*, ed. Oliver Mayo and Carolyn Leach (Adelaide, Australia: Wakefield, 2007): 39–47.

29. Nathaniel Comfort, *The Science of Human Perfection: How Genes Became the Heart of American Medicine* (New Haven, CT: Yale University Press, 2012), 207.

30. Lionel S. Penrose, "Phenylketonuria: A Problem in Eugenics," *Lancet* 1 (1946), 949–53, on 951.

31. The actual risk would, of course, be much higher than Penrose assumed. Given an incidence of 1 in 15,000, the carrier rate for the general population is about 1 in 60. Then, 2/3 × 1/60 chance of the mating = 2/180, or 1/90. For each offspring, the chance would then be 1/90 × 1/4 = 1/270. Neil R. M. Buist, personal communication, Nov. 22, 2011.

32. "The more practical remedy, advocating that carriers of the same defect should be discouraged from mating with one another, would efficiently diminish occurrence of abnormal homozygotes and would be eugenically acceptable." Lionel S. Penrose, *Outline of Human Genetics* (New York: John Wiley and Sons, 1959), 120.

33. Philosophers who have argued that "eugenics" per se is not repugnant include Philip Kitcher, in *The Lives to Come* (New York: Free Press, 1997); Allen Buchanan, Dan W. Brock, Norman Daniels, and Daniel Wikler, in *From Chance to Choice* (Cambridge: Cambridge University Press, 2000); Jonathan Glover, in *Choosing Children* (Oxford: Oxford University Press, 2006); and John Harris, in *Enhancing Evolution* (Princeton, NJ: Princeton University Press, 2007).

34. On Tay-Sachs and β-thalassemia screening, see Keith Wailoo and Stephen Pemberton, *The Troubled Dream of Genetic Medicine* (Baltimore: Johns Hopkins University Press, 2006); Ruth Schwartz Cowan, *Heredity and Hope: The Case for*

Genetic Screening (Cambridge, MA: Harvard University Press, 2008). For a detailed critical analysis of Tay-Sachs screening, see Aviad E. Raz, *Community Genetics and Genetic Alliances: Eugenics, Carrier Testing, and Networks of Risk* (London: Routledge, 2009). On screening in high schools, see Charles R. Scriver and John J. Mitchell, "Carrier Screening of Adolescents in Montreal," *eLS*, Dec. 2009, www.els.net/WileyCDA/ElsArticle/refId-a0005646.html; Lainie Friedman Ross, "Heterozygote Carrier Testing in High Schools Abroad: What Are the Lessons for the U.S.?" *Journal of Law, Medicine, and Ethics* 34, no. 4 (2006): 753–64.

35. Whonamedit? A Dictionary of Medical Eponyms, "Følling, Ivar Asbjørn," www.whonamedit.com/doctor.cfm/2400.html.

36. Penrose, "Memoirs," 8.

37. Deborah Clement Raessler, "Pearl S. Buck's Writings on Handicapped Children," in *The Several Worlds of Pearl S. Buck: Essays Presented at a Centennial Symposium, Randolph-Macon Woman's College, March 26–28, 1992*, ed. Elizabeth J. Lipscomb, Frances E. Webb, and Peter Conn (Westport, CT: Greenwood Press, 1994), 81–99, on 92.

Chapter 2. PKU as a Form of Cognitive Impairment

1. Philip M. Ferguson, "The Legacy of the Almshouse," in *Mental Retardation in America: A Historical Reader*, ed. Steven Noll and James Trent (New York: New York University Press, 2004), 48. See also Ferguson's *Abandoned to Their Fate: Social Policy and Practice toward Severely Retarded People in America, 1820–1920* (Philadelphia: Temple University Press, 1994).

2. James W. Trent, *Inventing the Feeble Mind: A History of Mental Retardation in the United States* (Berkeley: University of California Press, 1994), 40–59.

3. Peter L. Tyor and Leland V. Bell, *Caring for the Retarded in America: A History* (Westport, CT: Greenwood Press, 1984), esp. chap. 3.

4. One of Trent's key arguments in *Inventing the Feeble Mind* is that despite a broad trend from care to control, leaders of training schools mixed care and control from the time such schools were founded in the 1850s.

5. Penny Richards, "'Beside Her Sat Her Idiot Child': Families and Developmental Disability in Mid-Nineteenth-Century America," in Noll and Trent, *Mental Retardation*, 65–84.

6. Leila Zenderland, *Measuring Minds: Henry Herbert Goddard and the Origins of American Intelligence Testing* (Cambridge: Cambridge University Press, 1998), 75.

7. Direct quotes are from Parallels in Time: A History of Developmental Disabilities, "IV. The Rise of the Institutions 1800–1950," www.mnddc.org/parallels/four/4a/2.html. The Minnesota Governor's Council on Developmental Disabilities developed "Parallels in Time" (ca. 2000), an extensive multimedia website on the history of people with developmental disabilities. www.mnddc.org/parallels/index.html.

8. On cultural bias in tests used to detect morons, see Steven A. Gelb, "Social

Deviance and the 'Discovery' of the Moron," *Disability, Handicap, and Society* 2, no. 3 (1987): 247–58; Stephen Jay Gould, *The Mismeasure of Man* (New York: Norton, 1996).

9. Zenderland, *Measuring Minds*, chap. 6. See also Steven A. Gelb, "Goddard and the Immigrants, 1910–1917: The Studies and Their Social Context," *Journal of the History of the Behavioral Sciences* 22 (1986): 324–32.

10. Tyor and Bell, *Caring for the Retarded*, 118.

11. Molly Ladd-Taylor, "The 'Sociological Advantages' of Sterilization: Fiscal Policies and Feeble-Minded Women in Interwar Minnesota," in Noll and Trent, *Mental Retardation*, 281–99.

12. Molly Ladd-Taylor, "Who Is 'Defective' and Who Decides? The 'Feeble-minded' and the Courts" (paper presented at the American Association of the History of Medicine, Boston, May 2, 2003). Despite the widespread use of IQ tests by psychologists since the 1910s, their authority to diagnose mental retardation was contested well into the mid-twentieth century. Indeed, both the definition of MR and who decides which individuals have the condition have varied substantially over the past hundred years. For further discussion of this topic, see Jeffrey P. Brosco, "More Than the Names Have Changed: Exploring the Historical Epidemiology of Intellectual Disability in the US," in *Healing the World's Children: Comparative and Interdisciplinary Approaches to Child Health in the Twentieth Century*, ed. Cynthia Comacchio, Janet Golden, and George Weisz (Montreal: McGill-Queens University Press, 2008), 205–34.

13. Hamilton Cravens, "The Case of the Manufactured Morons: Science and Social Policy in Two Eras, 1934–1966," in *Technical Knowledge in American Culture: Science, Technology, and Medicine since the Early 1800s*, ed. Hamilton Cravens, Alan I. Marcus, and David M. Katzman (Tuscaloosa: University of Alabama Press, 1996), 151–68.

14. Steven A. Gelb, "'Mental Deficients' Fighting Fascism: The Unplanned Normalization of World War II," in Noll and Trent, *Mental Retardation*, 308–21.

15. Katherine Castles, "'Nice, Average Americans': Postwar Parents Groups and the Defense of the Normal Family," in Noll and Trent, *Mental Retardation*, 351–52.

16. Robert Segal, "The National Association for Retarded Citizens" (ca. 1974), www.thearc.org/page.aspx?pid=2342.

17. The Arc, Historical Accounts, "A History of the National Association for Retarded Children, Inc.," includes excerpts from *Blueprint for a Crusade: Publicity and Publications Manual* (1954), www.thearc.org/page.aspx?pid=2340.

18. The Arc, History of Research and Prevention, "1950s: A Decade of Decision," www.thearc.org/page.aspx?pid=2418.

19. The Arc, History of the Arc, "A History of Name Changes," www.thearc.org/page.aspx?pid=2344.

20. Castles, "Nice, Average Americans," 351–52.

21. Among the many historical works on the March of Dimes and the polio vac-

cine, see Jane Smith, *Patenting the Sun: Polio and the Salk Vaccine* (New York: Anchor/ Doubleday, 1991); David Rothman, *Beginnings Count: The Technological Imperative in American Health Care* (New York: Oxford University Press, 1997), 42–66.

22. Castles, "Nice, Average Americans," 352–53.

23. Quoted in ibid., 354–55.

24. Kathleen W. Jones, "Education for Children with Mental Retardation: Parent Activism, Public Policy, and Family Ideology in the 1950s," in Noll and Trent, *Mental Retardation*, 322–50.

25. Dale Evans Rogers, *Angel Unaware: A Touching Story of Love and Loss*, 50th anniversary ed. (Grand Rapids, MI: F. H. Revell, 2004); Trent, *Inventing the Feeble Mind*, 234.

Chapter 3. Testing and Treating Newborns, 1950–1962

1. Lionel S. Penrose, "Phenylketonuria: A Problem in Eugenics," *Lancet* 1 (1946): 949–53, on 951.

2. Ibid.

3. Lionel S. Penrose, *The Biology of Mental Defect*, preface by J. B. S. Haldane (London: Sidgwick and Jackson, 1949). This absence of discussion of diet contrasts with the 1954 edition, which was identical to the 1949 edition except for the correction of some errors and the addition of an appendix with a commentary on recent literature and additional references. The 1954 commentary mentions that many experiments aimed at the reduction of dietary phenylalanine had been conducted and that Bickel, Gerrard, and Hickmans had reported marked improvement in the mental state of a 2-year-old patient who was fed a phe-free diet. Lionel S. Penrose, *The Biology of Mental Defect*, rev. ed., preface by J. B. S. Haldane (London: Sidgwick and Jackson, 1954), appendix II, "Commentary on Recent Literature," 275–84, on 283.

4. Penrose was apparently influenced in this regard by the work of Harold Himwich and Joseph Fazekas, colleagues of George Jervis at Letchworth Village in New York. In a 1940 study based on analysis of blood samples drawn from patients at Letchworth, Himwich and Fazekas reported reduced utilization of oxygen (which they took to be a measure of cerebral metabolism) in "Mongolian idiocy" and PKU. They considered three ways in which diminished cerebral metabolism might be related to the mental deficiency associated with both diseases: (1) the diminished metabolism and associated structural changes early in life caused the mental deficiency; (2) they were independent effects of the same underlying cause, such as an enzyme deficiency; or (3) they were unrelated. Harold E. Himwich and Joseph F. Fazekas, "Cerebral Metabolism in Mongolian Idiocy and Phenylpyruvic Oligophrenia," *Archives of Neurology and Psychiatry* 44 (1040): 1213–18. (In the 1940s and 1950s, "phenylpyruvic oligophrenia," the name given to the disease by George Jervis, was still generally favored by researchers in the United States.) See Penrose, "Phenylketonuria," 952; Penrose, *Biology of Mental Defect* (1949), 145.

5. Louis I. Woolf, "The Early History of PKU" (unpublished typescript, Oct. 2007). Woolf also reiterated this claim in an interview with DBP in Vancouver, BC, June 15, 2009.

6. John W. Gerrard, "Phenylketonuria Revisited," *Clinical and Investigative Medicine* 17, no. 5 (1994): 510–13, on 512.

7. Louis I. Woolf and David G. Vulliamy, "Phenylketonuria with a Study of the Effect upon It of Glutamic Acid," *Archives of Diseases in Childhood* 26 (1951): 487–94.

8. Although Block is virtually never mentioned in connection with PKU, he is the same biochemist who collaborated with George Jervis and had unsuccessfully proposed development of a low-phe casein hydrolysate a decade before the idea occurred to Woolf.

9. Woolf, "Early History of PKU"; J. R. Alonso-Fernández and C. Colón, "The Contributions of Louis I Woolf to the Treatment, Early Diagnosis and Understanding of Phenylketonuria," *Journal of Medical Screening* 16, no. 4 (2009): 205–11.

10. Seymour Kaufman, *Overcoming a Bad Gene: The Story of the Discovery and Treatment of Phenylketonuria, a Genetic Disease That Causes Mental Retardation* (AuthorHouse, 2004), 49–50.

11. Woolf, "Early History of PKU."

12. Bickel wrote that he did so in an effort to impress his new colleague, John Gerrard. H. Bickel, "The First Treatment of Phenylketonuria," *European Journal of Pediatrics* 155, suppl. 1 (1996): S2–3, on S2.

13. Note the earlier effort with glutamic acid. Woolf and Vulliamy, "Phenylketonuria."

14. Jean Holt Koch, *Robert Guthrie—The PKU Story: A Crusade against Mental Retardation* (Pasadena, CA: Hope Publishing House, 1997), 23.

15. John W. Gerrard to Harvey L. Levy, May 31, 1977, Guthrie papers, box 22, "'PKU Controversy': 1966–1970."

16. New England Consortium of Metabolic Programs, "Discovery of the Diet for PKU by Dr. Horst Bickel," newenglandconsortium.org/for-professionals/video-library/discovery-of-the-diet-for-pku-by-dr-horst-bickel.

17. In 1946, the American Medical Association first asked physicians to obtain consent from patients before experimenting. For an introduction to the issues at the time, see Susan E. Lederer, *Subjected to Science: Human Experimentation in America before the Second World War* (Baltimore: Johns Hopkins University Press, 1995).

18. L. I. Woolf, Ruth Griffiths, and Alan Moncrieff, "Treatment of Phenylketonuria with a Diet Low in Phenylalanine," *British Medical Journal* 1 (1955): 57–64, on 63 (emphasis in original); Woolf, "Early History of PKU," 2.

19. L. I. Woolf, Ruth Griffiths, Alan Moncrieff, et al., "The Dietary Treatment of Phenylketonuria," *Archives of Disease in Childhood* 33 (1958): 31–45; Woolf, "Early History of PKU," 3.

20. The Dent group did not publish this result, which was privately commu-

nicated to Armstrong and other researchers. Marvin D. Armstrong and Frank H. Tyler, "Studies on Phenylketonuria: I. Restricted Phenylalanine Intake in Phenylketonuria," *Journal of Clinical Investigation* 34 (1955): 565–80.

21. Ibid., 578 and n. 21.

22. Frederick A. Horner and Charles W. Streamer, "Effect of a Phenylalanine-Restricted Diet on Patients with Phenylketonuria: Clinical Observations in Three Cases," *Journal of the American Medical Association* 161, no. 17 (1956): 1628–30.

23. Eugene A. Knox, "An Evaluation of the Treatment of Phenylketonuria with Diets Low in Phenylalanine," *Pediatrics* 26 (1960): 1–11.

24. Neil R. M. Buist, "The Evolution of Diets for Metabolic Disorders" (PowerPoint presentation, Phyllis Acosta Keynote Address, at the Genetic Metabolic Dieticians International conference, Emory University, Atlanta, Apr. 24, 2008).

25. George A. Jervis, "The Genetics of Phenylpyruvic Oligophrenia (A Contribution to the Study of the Influence of Heredity on Mental Defect)," *Journal of Mental Science* 85 (1939): 719–62.

26. The idea of using a drop of ferric chloride on a wet diaper was not original; in his first description of the College of Medical Evangelists' project, Centerwall noted that several writers had suggested trying this test if urine samples were unavailable. Willard R. Centerwall, "Phenylketonuria," *Journal of the American Medical Association* 165, no. 4 (1957): 392.

27. Koch, *Robert Guthrie*, 24–25.

28. Willard R. Centerwall, Robert F. Cinnock, and Albert Pusavat, "Phenylketonuria: Screening Programs and Testing Methods," *American Journal of Public Health* 50, no. 11 (1960): 1667–77; note also other methods.

29. Frank L. Lyman, "Preface," in *Phenylketonuria*, ed. Frank L. Lyman, foreword by Pearl S. Buck (Springfield, IL: Charles C. Thomas, 1963), xi–xii, on xi. This volume consists of papers presented at the First International Conference on Mental Retardation (July 1959).

30. Committee for the Study of Inborn Errors of Metabolism, National Research Council, *Genetic Screening: Programs, Principles, and Research* (Washington, DC: National Academy of Sciences, 1975), appendix H, "Screening for PKU in the United Kingdom," 347.

31. Ruth Brecher and Edward Brecher, "Saving Children from Mental Retardation," *Saturday Evening Post*, Nov. 21, 1959, 32–33, 109–11, on 108.

32. Kenneth A. Pass, "Lessons Learned from Newborn Screening for Phenylketonuria," in *Genetics and Public Health in the 21st Century: Using Genetic Information to Improve Health and Prevent Disease*, ed. Muin J. Khoury, Wylie Burke, and Elizabeth J. Thomson (New York: Oxford University Press, 2000), 385–404, on 385; Committee for the Study of Inborn Errors, *Genetic Screening*, 26. The Children's Bureau is now the Maternal and Child Health Bureau, Genetics Services Branch, Federal Health Resources and Services Administration.

33. "U.S. Panel Urges Testing at Birth: Acts to End Threat to Babies of Mental

Retardation," *New York Times*, Dec. 10, 1961, 80; Rudolph P. Hormuth, "Newborn Screening Systems" (proceedings of the 10th National Neonatal Screening Symposium, Seattle, WA, June 7–11,1994), 1–6.

34. "Projects Geared to Aid Retarded," *New York Times*, Apr. 7, 1957, 126.

35. M. Susan Lindee, *Moments of Truth in Genetic Medicine* (Baltimore: Johns Hopkins University Press, 2005), 32.

36. "Minutes, Technical Committee on Clinic Programs for Mentally Retarded Children," Sept. 11–12, 1958, USCB, NARA II, Central Files, 1958–62, box 981, file 1-2-3-26.

37. Pass, "Lessons Learned," 385.

38. Hormuth, "Newborn Screening Systems," 3.

39. Kaufman, *Overcoming a Bad Gene*, 54; Neil M. R. Buist, personal communication, Nov. 22, 2011.

40. Robert Guthrie, "Blood Screening for Phenylketonuria" (letter), *Journal of the American Medical Association* 178 (1961): 863; Robert Guthrie and Ada Susi, "A Simple Phenylalanine Method for Detecting Phenylketonuria in Large Populations of Newborn Infants," *Pediatrics* 32 (1963): 338–43.

41. Guthrie noted that the principle was identical to that of the bacterial assays he had used for cancer patients, except that in this case, the antimetabolite was in the agar and the metabolite in the blood. Robert Guthrie, "The Origin of Newborn Screening," *Screening* 1 (1992): 5–15, on 6.

42. Robert Guthrie, "Explorations in Prevention," in *Perspectives in Special Education: Personal Orientations*, ed. B. Blatt and R. Morris (Santa Monica, CA: Scott, Foresman, 1984), 157–72, on 163. For a profile of Margaret Doll, see Koch, *Robert Guthrie*, 155–59.

43. Blood serum lacks both red and white blood cells and clotting factors. With the use of whole blood, there was no need to centrifuge the sample to remove these cellular components.

44. Neil R. M. Buist notes that the phenylalanine content of plasma and red blood cells is almost identical, whereas for many other amino acids the content is very different, which explains why the Guthrie test worked so well in whole blood (though this was not understood at the time). Personal communication, Nov. 22, 2011.

45. Guthrie, "Origin of Newborn Screening," 7.

46. This story is told in detail in Koch, *Robert Guthrie*, 31–34. Kenneth Pass notes that, at the time the test was invented, new mothers typically stayed in the hospital for up to a week following delivery. Pass, "Lessons Learned," 368.

47. Edwin W. Naylor, interview, Isle of Pines, SC, Oct. 23, 2008.

48. Stanley W. Wright, "Editor's Column: Mass Screening for Phenylketonuria," *Journal of Pediatrics* 61 (Oct. 1962): 651–52.

49. Harvey L. Levy, "Historical Perspectives: Newborn Metabolic Screening," *NeoReviews* 6, no. 2 (2005): e57–60.

50. Guthrie, "Origin of Newborn Screening," 10.

51. Robert Guthrie, "Blood Screening for Phenylketonuria," *Journal of the American Medical Association* 178 (1961): 863.

52. Robert Guthrie and Ada Susi, "A Simple Phenylalanine Method for Detecting Phenylketonuria in Large Populations of Newborn Infants," *Pediatrics* 32 (1963): 338–42. Susi was hired by Guthrie in 1956 and would serve for 20 years as his chief technician. She and her husband, Karl, had been displaced from their native Estonia, where she had been a nurse. Although Ada Susi spent five years as a chief surgical nurse in Ohio, she was not allowed to work as a nurse in New York State because she was unable to provide records of her training in Estonia. Guthrie, "Explorations in Prevention," 160. A few further details about Susi can be found in Koch, *Robert Guthrie*. It is no longer possible to untangle Guthrie's and Susi's respective contributions. According to Ed Naylor, who had been a postdoctoral fellow with Guthrie, "she's the one that really developed the assay. [Following the conversation with Robert Warner] Bob went back to the lab and he told Ada Susi, see if we can develop an assay for phenylalanine. So Ada, she had been working with phenylalanine for a number of years and she, just in one day, came up with the assay." Edwin W. Naylor, interview, Isle of Pines, SC, Oct. 23, 2008. In a later communication, he elaborated: "The use of metabolic inhibitors for cancer chemotherapy was certainly Bob's idea and the use of beta-2-thienylalanine as a specific inhibitor that was reversed by phenylalanine was almost certainly Bob's. I'm almost certain that Bob gave her instructions to set up an assay for phenylalanine that could be used to quantitate phenylalanine levels for monitoring children with PKU. Ada then would have set up and validated the assay . . . The ideas were Bob's and the work in the lab was Ada's." Personal communication, Apr. 10, 2013.

Chapter 4. The Campaign for Mandatory Testing

1. Paul Starr, *The Social Transformation of American Medicine: The Rise of a Sovereign Profession and the Making of a Vast Industry* (New York: Basic Books, 1983).

2. G. E. R. Lloyd, ed., *Hippocratic Writings* (Harmondsworth, UK: Penguin Books, 1978).

3. Richard A. Meckel, *Save the Babies: American Public Health Reform and the Prevention of Infant Mortality* (Baltimore: Johns Hopkins University Press, 1990).

4. George Rosen, *Preventive Medicine in the United States, 1900–1975: Trends and Interpretations* (New York: Science History Publications, 1975).

5. Jeffrey P. Brosco, "The Early History of the Infant Mortality Rate in America: A Reflection upon the Past and a Prophecy of the Future," *Pediatrics* 103, no. 2 (1999): 478–85.

6. David J. Rothman, *Beginnings Count: The Technological Imperative in American Health Care* (New York: Oxford University Press, 1997).

7. Jane S. Smith, *Patenting the Sun: A History of the Salk Vaccine* (New York: William Morrow, 1990).

8. Allan M. Brandt, *No Magic Bullet: A Social History of Venereal Disease in the United States since 1980* (New York: Oxford University Press, 1985).

9. Charles E. Rosenberg, *The Cholera Years: The United States in 1832, 1849, and 1866* (Chicago: University of Chicago Press, 1962).

10. John Parascandola, *Sex, Sin, and Science: A History of Syphilis in America* (Santa Barbara, CA: Praeger, 2008).

11. Meckel, *Save the Babies*, 16.

12. On infant and child mortality, see Thomas McKeown, *The Modern Rise of Population* (London: Edward Arnold, 1976); J. B. McKinlay and S. M. McKinlay, "The Questionable Contribution of Medical Measures to the Decline of Mortality in the United States in the Twentieth Century," *Milbank Memorial Fund Quarterly: Health and Society* 55, no. 3 (1977): 405–28. On the mortality transition more generally, see Thomas McKeown, *The Role of Medicine: Dream, Mirage, or Nemesis* (Princeton, NJ: Princeton University Press, 1980), which expands on McKeown's classic 1955 article arguing that the decline in deaths due to infectious disease usually preceded the invention of effective medical methods or more and/or better hospitals. For commentaries on the controversial "McKeown thesis," see Bruce G. Link and Jo C. Phelan, "McKeown and the Idea That Social Conditions Are Fundamental Causes of Disease," *American Journal of Public Health* 92, no. 5 (2002): 722–25; James Colgrove, "The McKeown Thesis: A Historical Controversy and Its Enduring Influence," *American Journal of Public Health* 92, no. 5 (2002): 725–29; Simon Szreter, James Colgrove, Bruce G. Link, and Jo C. Phelan, "The McKeown Thesis," *American Journal of Public Health* 92, no. 5 (2002): 722–32; Bill Bynum, "The McKeown Thesis," *Lancet* 371, no. 9613 (2008): 644–45.

13. Paul J. Edelson, "History of Genetic Screening in the United States: I. The Public Debate over Phenylketonuria (PKU) Testing" (paper presented at the Annual Meeting of the American Association for the History of Medicine, New York, 1994).

14. Edward D. Berkowitz, "The Politics of Mental Retardation during the Kennedy Administration," *Social Science Quarterly* 61, no. 1 (1980): 128–43, on 140.

15. In a letter to Eunice Shriver, in which he listed the important groups working in the area of mental retardation, Robert E. Cooke wrote of the NARC that it had "contributed relatively little to research in the field of mental retardation in view of the size of the organization," although it had made major contributions to the development of improved services, especially those that would enable parents to care at home for their retarded child. Robert E. Cooke to Mrs. R. Sargent Shriver, Sept. 23, 1960, Joseph P. Kennedy, Jr. Foundation Archives, Washington, DC, box 265, folder 574.

16. Edelson, "History of Genetic Screening."

17. Masland continued promoting research in the etiology of mental retardation as director of the National Institute of Neurological Diseases and Blindness (an NIH institute) from 1959 to 1968.

18. Richard L. Masland, Seymour B. Sarason, and Thomas Gladwin, *Mental

Subnormality: Biological, Psychological, and Cultural Factors (New York: Basic Books, 1958), 50.

19. Edward Shorter, *The Kennedy Family and the Story of Mental Retardation* (Philadelphia: Temple University Press, 2000), 48.

20. *The Dark Corner* also became a prize-winning television documentary. In an oral history interview, Cooke noted that when the Shrivers first came to Baltimore to discuss the Johns Hopkins research program, they did not indicate any personal connection to the subject of MR, although their conversations gradually became more frank. Robert E. Cooke Oral History Interview, John F. Kennedy Oral History Collection, John F. Kennedy Presidential Library and Museum, Boston, MA, JFK #2, 7/25/1968, p. 31, www.jfklibrary.org/Asset-Viewer/Archives/JFKOH-REC-02.aspx. Cooke's efforts to change attitudes toward MR and his role in the making of the film are discussed in Alexandra Minna Stern, *Telling Genes: The Story of Genetic Counseling in America* (Baltimore: Johns Hopkins University Press, 2012), esp. 75–76.

21. Boggs recalled an incident in which a reference to President Kennedy's having a retarded sister in a caption to a photo in *Children Limited*, the NARC national newspaper, was soon followed by a bulletin to the local associations asking that no capital should be made of this fact, as the NARC should respect the family's wishes that it not be mentioned. She also noted that she was told by the head of the organization's Massachusetts chapter that it had tried to reach the family through channels in Boston but had received no response. Elizabeth M. Boggs Oral History Interview, John F. Kennedy Oral History Collection, John F. Kennedy Presidential Library and Museum, Boston, MA, JFK #1, 7/17/1968, pp. 4–5, www.jfklibrary.org/Asset-Viewer/Archives/JFKOH-EMB-01.aspx. Boggs provides a more oblique account of Kennedy family attitudes in "Federal Legislation Affecting the Mentally Retarded 1957–1967: An Historical Overview," in *Mental Retardation: An Annual Review*, vol. 3, ed. Joseph Wortis (New York: Grune and Stratton, 1971), 103–27, on 111–12.

22. Joe McCarthy, *The Remarkable Kennedys* (New York: Dial Press, 1960), 14; "President Spurs Retarded Study: Calls for National Program in Meeting New Panel," *New York Times*, Oct. 19, 1961, 24. Both book and article are noted in Gerald O'Brien, "Rosemary Kennedy: The Importance of a Historical Footnote," *Journal of Family History* 29, no. 3 (2004): 225–36, on 227. O'Brien also quotes David Koskoff's statement that "accounts of the family dismissed Rosemary as the quiet one who had elected to devote her life to helping the handicapped with the sisters of St. Coletta." David E. Koskoff, *Joseph P. Kennedy: His Life and Times* (Englewood Cliffs, NJ: Prentice-Hall, 1974), 335–36. At the behest of her father, Rosemary, who had become moody and sometimes aggressive, was subjected at age 23 to a prefrontal lobotomy. It proved disastrous, and she was eventually institutionalized at St. Coletta School for Exceptional Children in Wisconsin, where she resided until her death in 2005.

23. Eunice Kennedy Shriver, "Hope for Retarded Children" (*Saturday Evening Post*, 1962)," in *Mental Retardation in America: A Historical Reader*, ed. Steven Noll and James W. Trent (New York: New York University Press, 2004), 303–7.

24. The representative was Boggs, who would soon serve as NARC president. Boggs, Oral History Interview, 1.

25. Shorter, *Kennedy Family*, 74.

26. Deborah M. Spitalnik, *The President's Panel and the Public Policy Contributions of Eunice Kennedy Shriver* (New Brunswick, NJ: Elizabeth M. Boggs Center on Developmental Disabilities, 2011), 6.

27. "Statement by the President on the Need for a National Program to Combat Mental Retardation" (Oct. 11, 1961), www.jfklink.com/speeches/jfk/publicpapers/1961/jfk413_61.html.

28. President's Panel on Mental Retardation, *National Action to Combat Mental Retardation* (Washington, DC: Government Printing Office, 1962). Largely through the efforts of the president's sister Eunice Shriver, her spouse, Sargent Shriver, and pediatrician Robert Cooke, most of the panel's recommendations were implemented, including creation of the National Institute for Child Health and Human Development.

29. "A.M.A. Cites Gains by Medicine in '62," *New York Times*, Jan. 2, 1963, 1; Austin Wehrwein, "'Vast Medical Developments' Hailed," *Boston Herald*, Jan. 3, 1963, 2.

30. "Chromosomes and the Mind," *Time*, Dec. 14, 1962, 68–69.

31. Berkowitz, "Politics of Mental Retardation," 136.

32. Robert E. Cooke, "The Postnatal 'Prevention' and Correction of Congenital Defects," in *Second International Conference on Congenital Malformations* (New York: International Medical Congress, 1964), 386–93, on 392.

33. Masland et al., *Mental Subnormality*, 26.

34. Katherine B. Oettinger, Chief, US Children's Bureau, Press Release, Dec. 14, 1965, USCB, NARA II, Central Files, 1963–66.

35. Duane Alexander, "The National Institute of Child Health and Human Development and Phenylketonuria," *Pediatrics* 112, no. 6 (2003): 1514–15, on 1514.

36. "Needed: A Law to Prevent Heartbreak," *Parade* (Western edition), May 10, 1964. Mildred Plapinger Small, media consultant at the NARC, sent Guthrie a copy of the article with an accompanying note: "Guess it was worth the lunch I took the editor to!" Guthrie papers, box 1, folder "Press Clippings, 1858–1980."

37. Harold M. Schmeck, Jr., "Biochemical Detective Findings Lead to Gains in Mental Health," *New York Times*, May 21, 1961, 82. See also "For the Backward," *Newsweek*, Feb. 18, 1957, 102; Shirley Sirota Rosenberg, "A New Life for Karen," *Family Weekly*, Dec. 16, 1962, 6–7.

38. Gordon G. Greer, "Unnecessary Menace to Young Brains: PKU," *Better Homes and Gardens*, Feb. 1966, 25.

39. Harvey L. Levy, "Historical Perspectives: Newborn Metabolic Screening," *NeoReviews* 6, no. 2 (2005): e57–60.

40. President's Commission for the Study of Ethical Problems in Medicine and Biomedical and Behavioral Research, *Screening and Counseling for Genetic Conditions: A Report on the Ethical, Social, and Legal Implications of Genetic Screening, Counseling, and Education Programs* (Washington, DC: Government Printing Office, 1983), 13.

41. Robert Guthrie, "The Origin of Newborn Screening," *Screening* 1 (1992): 12.

42. Apart from the extreme rarity of homocystinuria, the BIA for the disease had other weaknesses, including high rates of false negatives and positives. Kenneth A. Pass, "Lessons Learned from Newborn Screening for Phenylketonuria," in *Genetics and Public Health in the 21st Century: Using Genetic Information to Improve Health and Prevent Disease*, ed. Muin J. Khoury, Wylie Burke, and Elizabeth J. Thomson (New York: Oxford University Press, 2000), 385–404, on 386–87.

43. The resolution was adopted on October 11, 1964, and published in the NARC's *Weekly Action Report*, no. 58 (Jan. 4, 1965). It is referenced in a memo of Robert M. Gettings, NARC Assistant for Governmental Affairs, to Wylie Bowmaster, Chairman, Legislative Committee, Tennessee ARC, July 29, 1966. The NARC also advised that no requirement as to the type of test should be written into the law and that testing of other types of metabolic disease should be permitted. Elizabeth M. Boggs Papers, Elizabeth M. Boggs Center on Developmental Disabilities, University of Medicine and Dentistry of New Jersey–Robert Wood Johnson Medical School, New Brunswick, NJ.

44. The NARC model bill was first circulated in October 1964. Committee for the Study of Inborn Errors of Metabolism, National Research Council, *Genetic Screening: Programs, Principles, and Research* (Washington, DC: National Academy of Sciences, 1975), 48.

45. Ibid., esp. "The Development of Legislation and Regulation for PKU Screening," 44–87.

46. Joseph D. Cooper, comments in "PKU Conference Essay" (digest of the taped proceedings of the Conference on PKU and Biomedical Legislation, Harvard University Program on Technology and Society, May 11, 1968, draft edited by Judith P. Swazey), 24. A copy of the transcript was provided by Harvey L. Levy, a conference participant.

47. Ibid., 27, 47.

48. *Parade*, "Needed."

49. "U.S. Panel Urges Testing at Birth: Acts to End Threat to Babies of Mental Retardation," *New York Times*, Dec. 10, 1961, 80. The article also cited the contrasting fates of the NARC's poster children, Kammy and Sheila McGrath, and asserted that, with the ferric chloride test, "diagnosis is simple."

50. "A Serious Threat to Children—PKU," *PTA Magazine* 59, no. 9 (1965): 24. The PTA advisory was quoted in Representative George P. Miller, "A Serious Threat to Children—PKU," Extension of Remarks of Hon. George P. Miller of California in the House of Representatives, Thursday, Aug. 26, 1965, *Congressional Record—Appendix*, Sept. 3, 1965, A5008–5010.

51. See, for example, President's Panel, *National Action*, 1–2, 47.

52. Montoya's comment in *Congressional Record*, 27919, Oct. 21, 1965; see also Ruth and Edward Brecher, "Saving Children from Mental Retardation," *Saturday Evening Post*, Nov. 21, 1959, 32–33, 109–11, on 32.

53. Samuel P. Bessman, Letter to the Editor, *New England Journal of Medicine* 273, no. 14 (1965): 772.

54. Children's Bureau, US Department of Health, Education, and Welfare, "An Inventory of Children with Phenylketonuria" (Nov. 1962), USCB, NCEMCH, www.mchlibrary.info/history/chbu/20363.pdf.

55. For example, Willard Centerwall estimated that each child institutionalized cost taxpayers $100,000. See "Phenylketonuria," *Currents in Public Health* (published by Ross Laboratories) 2, no. 2 (Feb. 1962): 3; Willard R. Centerwall, Robert F. Chinnock, and Albert Pusavat, "Phenylketonuria: Screening Programs and Testing Methods," *American Journal of Public Health and the Nation's Health* 50, no. 11 (1960): 1667–77, on 1667–68.

56. The expense of laboratory testing was equated with the unit cost of the test rather than with the cost of identifying one affected individual. (The latter would include the cost of retesting the large number of false positives intrinsic to screening for very rare conditions.)

57. For example, in Oregon, the loss was calculated at $5,000 per year per patient for an assumed 44 years of adult productive life. *Oregon Health Bulletin* 39 (May 1961): 1.

58. Harvey L. Levy, personal communication, June 19, 2012; Bradford L. Therrell, personal communication, June 6, 2012; Edwin W. Naylor, interview, Isle of Pines, SC, Oct. 23, 2008. For an overview of the history of NBS labs, see Bradford Therrell and John Adams, "Newborn Screening in North America," *Journal of Inherited Metabolic Disease* 30 (2007): 447–65.

59. Guthrie, "Origin of Newborn Screening," 8–10.

60. For more on the federal role in NBS, see Vince L Hutchins, *Maternal and Child Health at the Millennium* (Rockville, MD: Maternal and Child Health Bureau, 2001); Michele A Lloyd-Puryear, Bradford L. Therrell, Marie Mann, et al., "The Role of the Federal Government in Supporting State Newborn Screening Programs," in *Ethics and Newborn Genetic Screening: New Technologies, New Challenges*, ed. Mary Ann Baily and Thomas H. Murray (Baltimore, MD: Johns Hopkins University Press, 2009), 178–94; Jeffrey P. Brosco, "Navigating the Future through the Past: The Enduring Historical Legacy of Federal Children's Health Programs in the United States," *American Journal of Public Health* 102, no. 10 (2012): 1848–57, epub Aug. 16, 2012.

61. Levy, personal communication, June 19, 2012; Therrell, personal communication, June 6, 2012; Naylor, interview. See also Therrell and Adams, "Newborn Screening," 449–53.

Chapter 5. Sources of Skepticism

1. J. A. Anderson and K. F. Swaiman, eds., *Phenylketonuria and Allied Metabolic Diseases: Proceedings of a Conference Held at Washington, DC, April 6–8, 1966* (Washington, DC: Government Printing Office, 1967), 238.

2. Committee on the Handicapped Child of the American Academy of Pediatrics, "Statement on Phenylketonuria," Aug. 1964, USCB, NARA II, box 1000, file 4-5-11-5. This statement was not published, but a milder version appeared later. Committee on the Handicapped Child of the American Academy of Pediatrics, "Statement on Treatment of Phenylketonuria," *Pediatrics* 35 (Mar. 1965): 501–3. A statement by the AAP's Committee on Fetus and Newborn, "Screening of Newborn Infants for Metabolic Disease," *Pediatrics* 15 (Mar. 1965), 499–501, did not comment on legislation. However, a 1967 statement by the AAP opposed compulsory screening with a similar warning: "Failure to achieve expected health benefits as a result of premature and injudicious legislation may do irreparable harm to the orderly development of mass screening techniques for the early identification of disease and undermine public support of further research." American Academy of Pediatrics, "Statement on Compulsory Testing of Newborns for Metabolic Disorders," *Pediatrics* 39, no. 4 (1967): 84–85, on 84 (statement originally adopted at an AAP meeting of Sept. 5, 1966).

3. Samuel P. Bessman, "Some Biochemical Lessons to Be Learned from Phenylketonuria," *Journal of Pediatrics* 64, no. 6 (1964): 828–38, on 829.

4. Samuel P. Bessman, Raul Wapnir, and Dora Due, "Amine Metabolism in Phenylketonuria," in Anderson and Swaiman, *Phenylketonuria*, 19.

5. Samuel P. Bessman, "Implications of the Drive for Screening," in Anderson and Swaiman, *Phenylketonuria*, 177–80, on 179–80.

6. Julian L. Berman, George C. Cunningham, Robert W. Day, et al., "Causes for High Phenylalanine with Normal Tyrosine," *American Journal of Diseases of Childhood* 117 (Jan. 1969): 54–65, on 54.

7. Richard J. Allen, John C. Heffelfinger, Ronald E. Masotti, and Makepeace U. Tsau, "Phenylalanine Hydroxylase Activity in Newborn Infants," *Pediatrics* 33 (Apr. 1964): 512–25. See also David Yi-Yung Hsia, "Phenylketonuria: A Study of Human Biochemical Genetics, E. Mead Johnson Award Address, October 26, 1965," *Pediatrics* 38, no. 2, pt. 1 (1966): 173–84. This pattern of discovering the clinical complexity of a condition only *after* institution of widespread screening has complicated the expansion of NBS programs to many other conditions.

8. Helen K. Berry, Betty S. Sutherland, and Barbara Umbarger, "Phenylketonuria," *Disease-a-Month* 12, no. 12 (1966): 1–47, on 12. See also Helen K. Berry and Stanley Wright, "Conference on Treatment of Phenylketonuria," *Journal of Pediatrics* 70 (Jan. 1967): 142–47, on 144 (summary of a NARC-sponsored conference).

9. Committee for the Study of Inborn Errors of Metabolism, National Research Council, *Genetic Screening: Programs, Principles, and Research* (Washington, DC:

National Academy of Sciences, 1975), 28; Robert Guthrie and Ada Susi, "A Simple Phenylalanine Method for Detecting Phenylketonuria in Large Populations of Newborn Infants," *Pediatrics* 32 (1963): 338–43, on 342.

10. High blood phenylalanine levels can also result from mutations in genes that code for enzymes controlling the synthesis and regeneration of tetrahydrobiopterin (BH4), an essential cofactor for function of the PAH enzyme. For reasons of history and convenience, "phenylketonuria" has continued to designate what is now understood to be a heterogeneous entity characterized by both allelic and locus variation and a broad clinical spectrum.

11. California State Department of Public Health, Bureau of Maternal and Child Health, "Phenylketonuria and the Guthrie Inhibition Assay Screening Procedure, Summary of Meeting of Consultants to the California State Department of Public Health," *Pediatrics* 32 (1963): 345–46, on 346.

12. Paul J. Edelson, "Lessons from the History of Genetic Screening in the US: Policy Past, Present, and Future" (ms. prepared for the Hastings Center's project Priorities in the Clinical Application of Human Genome Research, 1995); Diane B. Paul and Paul J. Edelson, "The Struggle over Metabolic Screening," in *Molecularising Biology and Medicine: New Practices and Alliances, 1910s–1970s*, ed. Soraya de Chadarevian and Harmke Kamminga (Reading, UK: Harwood Academic Publishers, 1998), 203–20, on 214.

13. N. A. Holtzman, A. G. Meek, and E. D. Mellitis, "Neonatal Screening for Phenylketonuria: I. Effectiveness," *Journal of the American Medical Association* 229 (1974): 667–70.

14. David S. Kleinman, "Phenylketonuria: A Review of Some Deficits in Our Information," *Pediatrics* 33, no. 1 (1964): 123–34, on 125.

15. Pediatrician and bioethicist Norman Fost generated intense controversy when he charged in a 2005 *New York Times* interview that "if you give a normal kid a diet without enough phenylalanine, not only is there brain damage but every cell in the body is malnourished. Normal kids became brain-damaged. Many died." Gina Kolata, "Panel Advises Testing Babies for 29 Diseases," *New York Times*, Feb. 21, 2005. See also the summary of Norman Fost's presentation at the December 8, 2005, meeting of the President's Council on Bioethics, bioethics.georgetown.edu/pcbe/background/ethical_questions_screening.html. Brosco and colleagues have argued that, on the contrary, only a few cases of adverse outcomes arising from false-positive results can be documented and that it is unlikely (though not impossible) that publication bias explains their results. Jeffrey P. Brosco, Lee M. Sanders, Michael I. Seider, and Angela C. Dunn, "Adverse Medical Outcomes from Early Newborn Screening Programs for Phenylketonuria," *Pediatrics* 122 (2008): 192–97. Many contemporary statements do refer to cognitive damage and deaths but attribute them to dietary mismanagement rather than false-positive results; see, for example, American Academy of Pediatrics, Committee on the Handicapped Child, "Statement on Treatment of Phenylketonuria," *Pediatrics* 35 (1965): 501–3. "Over-

rigidity of dietary management has led to early death, presumably from insufficient protein intake or hypoglycemia" (503).

16. Mary L. Efron, Letter to the Editor, *New England Journal of Medicine* 273, no. 14 (1965): 772–73.

17. Berry et al., "Phenylketonuria," 13.

18. Edward R. B. McCabe and Linda McCabe, interview, Los Angeles, Feb. 22, 2007.

19. For example, a 1966 USCB-sponsored conference on the treatment of PKU stressed that the low-phe diet, far from being harmless, carried many potential risks. Berry and Wright, "Conference on Treatment," 144.

20. B. M. Rouse, "Phenylalanine Deficiency Syndrome," *Journal of Pediatrics* 69 (1966): 246–49; Allen et al., "Phenylalanine Hydroxylase Activity" (describing treatment of probable carriers). A conference paper abstract is referenced in Helen K. Berry, Betty S. Sutherland, and Barbara Umbarger, "Diagnosis and Treatment: Interpretation of Results of Blood Screening Studies for Detection of Phenylketonuria," *Pediatrics* 37 (1966): 102–6, ref. 5; and a personal communication in W. B. Hanley, L. Linsao, W. Davidson, and C. A. Moes, "Malnutrition with Early Treatment of Phenylketonuria," *Pediatric Research* 4, no. 4 (1970): 318–27.

21. Hanley et al., "Malnutrition."

22. William B. Hanley, interview, Toronto, Dec. 11, 2006.

23. Ellen Kang, "Comments Regarding Dr. Cooper's Letter to Dr. Hornig, November 23, Regarding PKU," attached to memo of Arthur J. Lesser, Deputy Chief, Children's Bureau, to Colin MacLeod, Office of Science and Technology, Executive Office of the President, Dec. 21, 1965, USCB, NARA II, box 1000, file 4-5-11-5.

24. Samuel P. Bessman, "Legislation and Advances in Medical Knowledge—Acceleration or Inhibition?" *Journal of Pediatrics* 69 (Aug. 1966): 334–38, on 336. Of the four cases reported, one was in England, one was in Israel, and two were the Massachusetts cases referenced by Mary Efron.

25. Shirley Sirota Rosenberg, "A New Life for Karen," *Family Weekly*, Dec. 16, 1962, 6–7, on 7.

26. Charles Johnson, Transcript of Proceedings, 7th PKU General Medical Conference, Vail, CO, Feb. 22–23, 1971, p. 181, Guthrie papers, box 15, binders "Collaborative Study of Children Treated for Phenylketonuria."

27. Virginia E. Schuett, Robert F. Gurda, and Eleanor S. Brown, "Diet Discontinuation Policies and Practices of PKU Clinics in the United States," *American Journal of Public Health* 70, no. 5 (1980): 498–503.

28. N. A. Holtzman, D. W. Welcher, and E. D. Mellits, "Termination of Restricted Diet in Children with Phenylketonuria: A Randomized Controlled Study," *New England Journal of Medicine* 293, no. 22 (1975): 1121–24.

29. Isabel Smith and Otto H. Wolff, "Duration of Treatment in Phenylketonuria," *Lancet* 303, no. 7868 (1974): 1229.

30. I. Smith, M. E. Lobascher, J. E. Stevenson, et al., "Effect of Stopping Low-

Phenylalanine Diet on Intellectual Progress of Children with Phenylketonuria," *British Medical Journal* 2, no. 6139 (1978): 723–26.

31. N. A. Holtzman, R. A. Kronmal, W. van Doorninck, et al., "Effect of Age of Loss of Dietary Control on Intellectual Performance and Behavior of Children with Phenylketonuria," *New England Journal of Medicine* 314 (1986): 593–98; C. Azen, R. Koch, E. Friedman, et al., "Summary of Findings from the United States Collaborative Study of Children Treated for Phenylketonuria," *European Journal of Pediatrics* 155, suppl. 1 (1996): S29–32. School history data were especially worrying, with about a third of children experiencing significant problems, such as repeating a grade or attending special classes (S31).

32. Stanley High, "Forgotten Children No Longer," *Reader's Digest*, Sept. 1960, 120–24, on 123. In a piece on the Kennedy Foundation's award to Følling, *Time* magazine reported that "now the defect can be promptly detected, and children started on a special diet escape nearly all the brain damage." "Chromosomes and the Mind," *Time*, Dec. 1962, 68–69, on 68. See also "Detecting Poisons at Birth," *Time*, Aug. 23, 1963, 47.

33. Berry et al., "Phenylketonuria," 8.

34. Ibid., 45.

35. Berry and Wright, "Conference on Treatment," 146.

36. Committee for the Study of Inborn Errors, *Genetic Screening*, 28.

37. At a 1968 conference, where he had argued against the view that high phenylalanine levels were causally related to mental retardation, Bessman stated that his group had been treating a child with tyrosine alone and that "the child has developed normally physically, and has not had convulsions or eczema." Quoted in "PKU Conference Essay" (digest of the taped proceedings of the Conference on PKU and Biomedical Legislation, Harvard University Program on Technology and Society, May 11, 1968, draft edited by Judith P. Swazey), 4. A copy of the transcript was provided by Harvey L. Levy, a conference participant. No report of the case was published until 1981, a decade after the end of the four-year experiment, apparently at the insistence of one of the coauthors who had inherited Bessman's patients in Maryland. Mark L. Batshaw, David Valle, and Samuel P. Bessman, "Unsuccessful Treatment of Phenylketonuria with Tyrosine," *Journal of Pediatrics* 99, no. 1 (1981): 159–60. According to Bessman, "the parents noted the child to be irritable and lethargic and requested that the [low phenylalanine] diet be terminated" (159), asking whether another type of therapy might be available. In the 1970s, Bessman speculated that the damage from tyrosine deficiency occurred before birth and was irreversible.

38. Samuel P. Bessman, "Genetic Failure of Fetal Amino Acid 'Justification': A Common Basis for Many Forms of Metabolic, Nutritional, and 'Nonspecific' Mental Retardation," *Journal of Pediatrics* 81, no. 4 (1972): 834–42; Samuel P. Bessman, Malcolm L. Williamson, and Richard Koch, "Diet, Genetics, and Mental Retardation Interaction between Phenylketonuric Heterozygous Mother and Fetus

to Produce Nonspecific Diminution of IQ: Evidence in Support of the Justification Hypothesis," *Proceedings of the National Academy of Sciences* 75, no. 3 (1978): 1562–66; Samuel P. Bessman, "The Justification Theory: The Essential Nature of the Non-Essential Amino Acids," *Nutrition Reviews* 37, no. 7 (1979): 209–20. Bessman's thesis was criticized in Charles R. Scriver, David E. C. Cole, Sally A. Houghton, et al., "Cord-Blood Tyrosine Levels in the Full-Term Phenylketonuric Fetus and the 'Justification Hypothesis,'" *Proceedings of the National Academy of Sciences* 77, no. 10 (1980): 6175–78.

39. Deposition of Samuel P. Bessman in *Gemeinhardt and Gemeinhardt vs. Colorado General Hospital*, Dec. 13, 1976, 16; see also Bessman's deposition in *Martha Cramer Lewis vs. H. Leo Owen, MD*, p. 37; both in Guthrie papers, box 3, subject files A–Z, 1960–1991, A-CAL.

40. Neil A. Holtzman, "Anatomy of a Trial," *Pediatrics* 60 (1977): 932–34.

41. Bessman, "Legislation and Advances," 337. Because Waldo Nelson, editor of the *Journal of Pediatrics*, shared this concern, he published Bessman's commentary as an Editor's Column.

42. "Minutes of Children's Bureau Ad Hoc Committee on Medical Genetics," Mar. 11, 1966, USCB, NARA II, box 995, file 4-4-1.

43. C. Azen, R. Koch, E. Friedman, et al., "Summary of Findings from the United States Collaborative Study of Children Treated for Phenylketonuria," *European Journal of Pediatrics* 155, suppl. 1 (1996): S29–32.

44. Mary L. Efron, Dean Young, Hugo W. Moser, and Robert A. MacCready, "A Simple Chromatographic Screening Test for the Detection of Disorders of Amino Acid Metabolism," *New England Journal of Medicine* 270, no. 26 (1964): 1378–83. Efron had learned chromatographic techniques while working with Charles Dent at University College London in the late 1950s. See Roland Westall, "Tributes: Mary Efron (Nov 18, 1926 to Sept 2, 1967)," *American Journal of Diseases of Childhood* 117 (Jan. 1969): 1–2 (Mary Efron Memorial Issue).

45. Efron, Letter to the Editor.

46. Herbert Black, "AMA Rejects Bay State's Infant Tests," *Boston Globe*, Dec. 3, 1964, 21. The AMA's Committee on Maternal and Child Care and Council on Medical Service deemed that "compulsory testing for PKU would be ill-advised since this might set a precedent for compulsory testing of newborn infants in hospitals for many other conditions." "Report of the Council on Medical Service: Voluntary Testing of Infants for PKU" (1964), personal papers of Shirley Sirota Rosenberg, Washington, DC.

47. Efron, Letter to the Editor.

48. According to contemporaries, Efron was an ardent admirer of Ayn Rand, and in fact she was the leading figure in the Rand group in Boston. Harvey L. Levy, personal communication, May 9, 2012; Mary Ampola, personal communication, May 15, 2012. See also "Nathaniel Branden to Give Lecture on Novelist Ayn Rand's Philosophy," *The Tech*, Oct. 16, 1963, 7, tech.mit.edu/archives/VOL_083/TECH_

V083_S0251_P007.pdf. Ampola believes that Efron's opposition to compulsion was also motivated by the fact that in Massachusetts, virtually every child was being tested before the law was passed, simply because it was good medical practice. Efron assumed that other states would follow and thus there was no need for government involvement.

49. Another 33 percent believed that the costs outweighed the benefits, while 41 percent had no opinion. (Benefits from screening for PKU were perceived to be even lower than for sickle-cell disease!) Moreover, only one-third of primary care physicians in the United States supported a "PKU registry" to keep track of patients. Irwin M. Rosenstock, Barton Childs, and Artemis P. Simopoulos, *Genetic Screening: A Study of the Knowledge and Attitudes of Physicians* (Washington, DC: National Academy of Sciences, 1975), table 23, p. 29, and pp. 41–42. The dim view of benefit would seem to be partly but not fully explained by physicians' opposition to legal mandates (see table 57, p. 54). Survey results were summarized in Committee for the Study of Inborn Errors, *Genetic Screening*, appendix G, "Data on Physicians' Knowledge and Attitudes," table G-18, p. 332. See also Barton Childs, "Persistent Echoes of the Nature-Nurture Argument," *American Journal of Human Genetics* 29 (1977): 1–13.

50. N. R. M. Buist and B. M. Jhaveri, "A Guide to Screening Newborn Infants for Inborn Errors of Metabolism," *Journal of Pediatrics* 82, no. 3 (1973): 511–22, on 514.

51. Berry et al., "Phenylketonuria," 12–13.

52. Bessman developed the first effective treatment for acute lead poisoning and did work on the chemical causes of hepatic coma that revolutionized its prevention and treatment. Among the six medical devices that he invented and patented was the first glucose sensor. "In Memoriam: Samuel P. Bessman, 89," *USC News*, Jan. 24, 2011, news.usc.edu/#!/article/30112/In-Memoriam-Samuel-P-Bessman-89. See also Justia Patents, "Patents by Inventor Samuel P. Bessman," http://patents.justia.com/inventor/SAMUELPBESSMAN.html.

53. See, for example, Samuel P. Bessman, Letter to the Editor, *New England Journal of Medicine* 273, no. 14 (1965): 772, in which he characterizes AMA opposition as "a petulant insistence that the Government have nothing to do with medicine."

54. Samuel P. Bessman and Judith P. Swazey, "Phenylketonuria: A Study of Biomedical Legislation," in *Human Aspects of Biomedical Innovation*, ed. Everett Mendelsohn, Judith P. Swazey, and Irene Taviss (Cambridge, MA: Harvard University Press, 1971), 49–78, on 51.

55. Rob Warden, "PKU Test Valid?" *Chicago Daily News*, Apr. 17, 1967, sect. 2, p. 21. In another episode, Dr. Ronald Davidson, director of the Division of Human Genetics at Children's Hospital Buffalo—Guthrie's home institution—agreed with Bessman's criticism of newborn screening for PKU. Davidson later requested that he not be quoted at all in relation to the PKU controversy. "Criticism Is Overstated,

Buffalo Doctor Contends," *Niagara Falls Gazette*, Jan. 16, 1969, Guthrie papers, box 1, "Press Clippings 1958–1980."

56. American Academy of Pediatrics, Committee on Nutrition, "Nutritional Management in Hereditary Metabolic Disease," *Pediatrics* 40, no. 2 (1967): 289–304, on 289.

57. "New Senate Bill Would Extend PKU Testing," *Medical Tribune and Medical News* 6, no. 10 (1965): 8.

58. Elizabeth Boggs, "Notes on Opposition to PKU Legislation," undated but probably written in late 1965 or early 1966. Boggs expressed similar reservations in the memos "Federal PKU Legislation," Sept. 9, 1965; "HR-11794" (referring to a parallel bill of Jan. 13, 1966, introduced in the House by John E. Moss, D-CA); and "Federal PKU Bills," Feb. 18, 1966; all in Elizabeth M. Boggs Papers, Elizabeth M. Boggs Center on Developmental Disabilities, University of Medicine and Dentistry of New Jersey–Robert Wood Johnson Medical School, New Brunswick, NJ, subject section "Phenylketonuria (PKU)."

59. Deborah M. Spitalnik, *The President's Panel and the Public Policy Contributions of Eunice Kennedy Shriver* (New Brunswick, NJ: Elizabeth M. Boggs Center on Developmental Disabilities, 2011), 9.

60. Gunnar Dybwad, "Farewell Address" (presented at the 1963 Annual Convention of the National Association for Retarded Children, Washington, DC, Oct. 26, 1963), www.disabilitymuseum.org/dhm/lib/detail.html?id=2233&&page=4.

61. Joseph D. Cooper, "Problems of Legislation in the Field of Mental Retardation," Scientific Seminars of the Rosewood State Hospital, Owings Mills, Maryland, Nov. 21, 1965, 12–13, USCB, NARA II, box 1000, file 4-5-11-5. See also Joseph D. Cooper, "More Problems," *Saturday Review* 50 (June 3, 1967): 56–61.

62. Robert A. MacCready, "PKU Testing: A Reply to Professor Cooper," *Medical Tribune and Medical News* 7, no. 42 (1966): 15.

63. Robert Guthrie, draft letter in response to article by Joseph D. Cooper, "Legislative Templates for Medical Practice: Errors of the Simplistic Model," *Medical Tribune and Medical News* 7, no. 9 (1966): 1, 15. Guthrie shared the draft with Ellen Kang of the Children's Bureau, who worried that "Cooper's reply to a letter by Guthrie would be that the Bureau put Guthrie up to this and is merely trying to sidestep the issue." Kang and Guthrie agreed that the draft reply would be sent to the NARC Public Health Committee, and the committee's chair ultimately appeared as the author. MacCready, "PKU Testing," 15. Guthrie's draft response, a memo from Rudolf P. Hormuth to Dr. Arthur J. Lesser of Feb. 18, 1966, describing the discussions with Kang, and copies of Cooper's and MacCready's articles are in USCB, NCEMCH, box 071653.

64. Norman Daniels surveys the reasons that identified victims are favored and analyzes the moral force of this bias in "Reasonable Disagreement about Identified vs. Statistical Victims," *Hastings Center Report* 42, no. 1 (2012): 35–45.

65. American Academy of Pediatrics, "Statement on Compulsory Testing," 84.

The editors of a USCB-supported conference publication on PKU similarly criticized "legislation which has included public pressure for a specific therapy at a time when competent medical investigators are not in agreement on the criteria for diagnosis, natural course of the disease or optimal therapy or therapies." Anderson and Swaiman, *Phenylketonuria*, 238. See also Berry and Wright, "Conference on Treatment."

66. Bessman, "PKU Conference Essay," 24.

67. MacCready quoted in "MacCready Defends Test," *Children Limited*, Feb. 1967, 4, Guthrie papers, box 17, folder "Feb. '67 Children Limited." The story "Guthrie Supports Mandatory Laws" in the same issue of *Children Limited* quotes Guthrie's charge that Bessman and Joseph Cooper confused the issues of testing and treatment.

68. Charles R. Scriver notes that, in contrast to the situation in the United States, the Quebec Network of Genetic Medicine could, from the start, integrate screening, confirmatory diagnostics, education, and treatment under the province's universal health insurance. Interview, Montreal, Nov. 13, 2008.

69. Robert Guthrie, "Screening of Newborn Infants for Metabolic Disorders: The Need for Federal Assistance for a Regional Approach" (1978), 1–2, included with memo of Mar. 29, 1978, to Dr. Jean Lockhart et al., Guthrie papers, box 2, folder "American Academy of Pediatrics."

Chapter 6. New Paradigms for PKU

1. Lionel S. Penrose, "Phenylketonuria: A Problem in Eugenics," *Lancet* 1 (1946): 949–53.

2. Charles R. Scriver, interview, Montreal, Nov. 13, 2008. See also Charles R. Scriver Interview, First Session—August 22, 2006, UCLA Oral History of Human Genetics Project, 65, ohhgp.pendari.com/Interview.aspx?id=33; Charles R. Scriver, "The PAH Gene, Phenylketonuria, and a Paradigm Shift," *Human Mutation* 28 (2007): 831–45.

3. David J. Rothman, *Beginnings Count: The Technological Imperative in American Health Care* (New York: Oxford University Press, 1997).

4. The terms "genetic" and "hereditary" are sometimes used interchangeably, but the difference is important in some cases. Traits may be considered genetic (by some definitions) but not hereditary, and the converse is also true. Chromosomal abnormalities such as Down syndrome, for example, are "genetic," but they usually result from a new mutation during the formation of eggs or sperm just before conception and so are not "inherited" from a previous generation. Conversely, traits can be hereditary but not involve changes in the sequence of bases in the DNA, instead involving other chemical changes in the DNA (so-called epigenetic changes) that affect gene expression.

5. At that time, abnormal mucus accumulation was taken to be the defining feature of cystic fibrosis. Keith Wailoo and Stephen Pemberton, *The Troubled Dream*

of Genetic Medicine: Ethnicity and Innovation in Tay-Sachs, Cystic Fibrosis, and Sickle Cell Disease (Baltimore: Johns Hopkins University Press, 2006), 74.

6. Committee for the Study of Inborn Errors of Metabolism, National Research Council, *Genetic Screening: Programs, Principles, and Research* (Washington, DC: National Academy of Sciences, 1975), 51, 92, on 92.

7. Ibid., 44.

8. Robert L. Sinsheimer, "The Prospect of Designed Genetic Change," *Engineering and Science* 13 (1969): 8–13.

9. D. S. Frederickson, *The Recombinant DNA Controversy: A Memoir* (Herndon, VA: ASM Press, 2001).

10. On the rDNA controversy, see Susan Wright, *Molecular Politics: Developing American and British Regulatory Policy for Genetic Engineering, 1972–1982* (Chicago: University of Chicago Press, 1994); Herbert Gottweis, *Governing Molecules: The Discursive Politics of Genetic Engineering in Europe and the United States* (Cambridge, MA; MIT Press, 1998); Frederickson, *Recombinant DNA Controversy*; M. J. Peterson, "Asilomar Conference on Laboratory Precautions," *International Dimensions of Ethics Education in Science and Engineering* (2010), www.umass.edu/sts/ethics.

11. The comment is from Richard Nixon's statement on signing the bill. The American Presidency Project, "Richard Nixon," www.presidency.ucsb.edu/ws/index.php?pid=3413#axzz1UYYbiwS8. The Black Panthers developed a rival program, offering their own sickle-cell testing service in many cities. Ruth Schwartz Cowan, *Heredity and Hope: The Case for Genetic Screening* (Cambridge, MA: Harvard University Press, 2008), 175–78.

12. Committee for the Study of Inborn Errors, *Genetic Screening*, iv.

13. Ibid., 88–93, 195, on 93; Samuel P. Bessman, "Legislation and Advances in Medical Knowledge—Acceleration or Inhibition?" *Journal of Pediatrics* 69 (Aug. 1966): 334–38.

14. Barron H. Lerner, "Beyond Informed Consent: Did Cancer Patients Challenge Their Physicians in the Post–World War II Era?" *Journal of the History of Medicine and Allied Sciences* 59 (2004): 507–21.

15. Gregory E. Pence, "Comas: Karen Quinlan and Nancy Cruzan," in *Classic Cases in Medical Ethics: Accounts of Cases That Have Shaped Medical Ethics, with Philosophical, Legal, and Historical Backgrounds* (New York: McGraw-Hill, 1995), 29–57. In the context of clinical research, the need for specific consent was established in the 1970s, in the aftermath of the Willowbrook, Tuskegee, and other scandals involving experiments with human subjects. *Advisory Committee on Human Radiation Experiments—Final Report* (New York: Oxford University Press, 1996), 80, 97–103, http://archive.org/details/advisorycommitte00unit; see also useful supplementary material at www.gwu.edu/~nsarchiv/radiation.

16. Committee for the Study of Inborn Errors, *Genetic Screening*, 121–24, 127.

17. Public Law 94-278 (1976), Title IV, 403(a), 90 Stat. 408.

18. Committee for the Study of Inborn Errors, *Genetic Screening*, 49–50, 93, on 93.

19. Ibid., 106, 188–89, on 106.

20. Apart from their use in developing new NBS tests, dried blood spots can be useful for many other kinds of research, including cancer genetics and assessment of environmental exposures during pregnancy. See epilogue for a discussion of issues regarding research with retained dried blood spots.

21. L. B. Andrews, J. E. Fullarton, N. A. Holtzman, and A. G. Motulsky, eds., *Assessing Genetic Risks: Implications for Health and Social Policy* (Washington, DC: National Academies Press, 1994), 51.

22. National Human Genome Research Institute, "Genetic Information Nondiscrimination Act (GINA) of 2008," www.genome.gov/24519851.

23. For an expanded version of this argument, see Diane B. Paul, "What Is a Genetic Test and Why Does It Matter?" *Endeavour* 23 (1999): 159–61.

24. For a good review of the arguments for informed consent for NBS, see Lainie Friedman Ross, "Mandatory versus Voluntary Consent for Newborn Screening," *Kennedy Institute of Ethics Journal* 20 (2011): 299–328.

25. Committee for the Study of Inborn Errors, *Genetic Screening*, 195.

26. Arthur Jensen, "How Much Can We Boost IQ and Scholastic Achievement?" *Harvard Educational Review* 39 (1969): 1–123.

27. Richard J. Herrnstein, "I.Q.," *Atlantic* 228 (1971): 63–64; Richard J. Herrnstein, *I.Q. in the Meritocracy* (Boston: Little Brown, 1973).

28. Leon Kamin, *The Science and Politics of IQ* (Potomac, MD: Lawrence Erlbaum, 1974), 1.

29. Evelyn Fox Keller has recently argued that the variables in IQ are so entangled that the question of how much observed differences in IQ are due to differences in genes and how much to differences in environment is simply unanswerable. Evelyn Fox Keller, *The Mirage of a Space between Nature and Nurture* (Durham, NC: Duke University Press, 2010), 53–54.

30. Ned J. Block and Gerald Dworkin, "IQ, Heritability, and Inequality," in *The IQ Controversy*, ed. Ned J. Block and Gerald Dworkin (New York: Pantheon, 1976), 410–542, on 489. See also Steven Rose, "Environmental Effects on Brain and Behaviour," in *Race, Culture, and Intelligence*, ed. Ken Richardson and David Spears (Baltimore: Penguin Books, 1972), 135.

31. Edward O. Wilson, *Sociobiology: The New Synthesis* (Cambridge, MA: Harvard University Press, 1975).

32. Philip Kitcher, *Vaulting Ambition: Sociobiology and the Quest for Human Nature* (Cambridge, MA: MIT Press, 1985), 128. Martin Barker similarly writes that "the significant thing is that they [genetic and other biological constraints] are capable of investigation, adaptation and alteration. This is once again the significance of PKU . . . To the extent that we conquer for a child its tendency to brain damage, we enable it to live by its potential for adaptability." Martin Barker, "Biology and Ideology: The Uses of Reductionism," in *Against Biological Determinism*, Dialectics of Biology Group, general ed. Steven Rose (London: Allison and Busby, 1982), 9–29, on 24.

33. Richard J. Herrnstein and Charles Murray, *The Bell Curve: Intelligence and Class Structure in American Life* (New York: Free Press, 1994); Robert Wright, "Dumb Bell," *New Republic*, Jan. 2, 1995, 6.

34. Leroy Hood, "The Book of Life" (commencement address, Whitman College, May 19, 2002), www.systemsbiology.org/download/whitman.pdf.

35. Walter Gilbert, "A Vision of the Grail," in *The Code of Codes: Scientific and Social Issues in the Human Genome Project*, ed. Daniel J. Kevles and Leroy Hood (Cambridge, MA: Harvard University Press, 1992), 83–97, on 94. For a similar comment, see Leroy Hood, "Biology and Medicine in the Twenty-First Century," in Kevles and Hood, *Code of Codes*, 112–63, on 138.

36. James D. Watson, "A Personal View of the Project," in Kevles and Hood, *Code of Codes*, 164–73, on 167. For more on the US genome project, see Charles DiLisi, "Meetings That Changed the World: Santa Fe 1986: Human Genome Baby-Steps," *Nature* 455 (Oct. 16, 2008): 876–77, epub Oct. 15, 2008, doi:10.1038/455876a; Robert Cook-Deegan, *The Gene Wars: Science, Politics, and the Human Genome* (New York: W. W. Norton, 1994); V. K. McElheny, *Drawing the Map of Life: Inside the Human Genome Project* (New York: Basic Books, 2010).

37. Richard C. Lewontin, "The Dream of the Human Genome," *New York Review of Books*, May 28, 1992.

38. Henry N. Kirkman, "Projections of a Rebound in Frequency of Mental Retardation from Phenylketonuria," *Applied Research in Mental Retardation* 3 (1982): 319–28, on 326.

39. Callers' comments and Robert Waterston's reply on the program "Human Genome Project," *Talk of the Nation*, hosted by Ira Flatow, National Public Radio, Nov. 15, 1996, transcript, 11–12. Other guests were Roger Brent and David Lipman.

40. Francis Collins, interview with Noah Adams, *All Things Considered*, National Public Radio, Dec. 2, 1999, transcript.

41. Angus J. Clarke, "Newborn Screening," in *Genetics, Society, and Clinical Practice*, ed. Peter S. Harper and Angus J. Clarke (Oxford: Bios Scientific Publishers, 1997), 107–17, on 107.

42. Daniel L. Hartl and Elizabeth W. Jones, *Genetics: Analysis of Genes and Genomes*, 7th ed. (Sudbury, MA: Jones and Bartlett, 2008), 2.

43. John J. Mitchell, "Phenylalanine Hydroxylase Deficiency," *GeneReviews*, Jan. 10, 2000, updated May 4, 2010, www.ncbi.nlm.nih.gov/books/NBK1504.

44. Seymour Kaufman, *Overcoming a Bad Gene* (AuthorHouse, 2004), 77–79.

45. Sapropterin dihydrochloride (Kuvan) mimics tetrahydrobiopterin (BH4), a coenzyme required by phenylalanine hydroxylase, and can reduce blood phenylalanine levels in people with hyperphenylalaninemia due to BH4-responsive PKU. A low-phe diet is still necessary, but treatment with Kuvan may allow some people to ingest more protein. At a National Institutes of Health PKU review conference in 2012, experts and PKU advocates debated the value of genotyping people for PKU to determine whether they are likely to benefit from the expensive drug. National

Institutes of Health, Eunice Kennedy Shriver National Institute of Child Health and Human Development (NICHD), Office of Rare Diseases Research (ORDR), and Office of Dietary Supplements (ODS), "Phenylketonuria Scientific Review Conference: State of the Science and Future Research Needs" (Natcher Conference Center, NIH, Bethesda, MD, 2012).

46. Chris Feudtner, *Bittersweet: Diabetes, Insulin, and the Transformation of Illness* (Chapel Hill: University of North Carolina Press, 2003). There are two forms of diabetes. Type 1 (the subject of Feudtner's book), often also known as juvenile or insulin-dependent diabetes, ordinarily manifests in childhood or adolescence. It requires careful monitoring of blood glucose levels and carbohydrate intake and multiple administrations of insulin daily, usually by injection. Type 2, or non-insulin-dependent diabetes, is much more common but typically milder. Its onset is generally after the age of 40, although it is being increasingly diagnosed in children.

Chapter 7. Living with PKU

1. N. A. Holtzman, D. W. Welcher, and E. D. Mellits, "Termination of Restricted Diet in Children with Phenylketonuria: A Randomized Controlled Study," *New England Journal of Medicine* 22 (1975): 1121–24.

2. National Institutes of Health, "Phenylketonuria: Screening and Management," Consensus Development Conference Statement (Oct. 16–18, 2000), consensus.nih.gov/2000/2000Phenylketonuria113html.htm; National Institutes of Health, Eunice Kennedy Shriver National Institute of Child Health and Human Development (NICHD), Office of Rare Diseases Research (ORDR), and Office of Dietary Supplements (ODS), "Phenylketonuria Scientific Review Conference: State of the Science and Future Research Needs" (Natcher Conference Center, NIH, Bethesda, MD, 2012). The authors of a recent Cochrane report note, however, that even the few studies on discontinuation and relaxation of the phe-restricted diet that met their standards for inclusion in the analysis were characterized by small numbers of participants, lack of follow-up into adulthood, and other methodological weaknesses. They conclude that "there is a lack of clear evidence about the precise level of phenylalanine restriction or when, if ever, the restricted diet should be relaxed" and that a large, well-designed, multicenter, randomized controlled trial is needed to resolve the uncertainties. V. J. Poustie and J. Wildgoose, "Dietary Interventions for Phenylketonuria (Review)," *Cochrane Database Systematic Review* 1 (2010), 7. For another skeptical view (of this and several other accepted principles), see W. B. Hanley, "Phenylketonuria: Questioning the Gospel," *Expert Review of Endocrinology and Metabolism* 2, no. 6 (2007): 809–16.

3. A. MacDonald, J. C. Rocha, M. van Rijn, and F. Feillet, "Nutrition in Phenylketonuria," *Molecular Genetics and Metabolism* 104 (2011): S10–18, on S11.

4. The term "medical food" is sometimes also applied to foods modified to be low in protein. Terminological confusions and their consequences are discussed in Kathryn M. Camp, Michele A. Lloyd-Puryear, and Kathleen L. Huntington, "Nu-

tritional Treatment for Inborn Errors of Metabolism: Indications, Regulations, and Availability of Medical Foods and Dietary Supplements using Phenylketonuria as an Example," *Molecular Genetics and Metabolism* 107 (2012): 3–9.

5. In the United States, target ranges vary with clinic. A common recommendation is 2 to 6 mg/dl (120–360 μmol/L). See National PKU Alliance, "Chapter Three: Monitoring Blood Phenylalanine Levels," in *The PKU Handbook: A Guide to PKU from Diagnosis to Adult* (Tomahawk, WI: National PKU Alliance, 2011), 26–28, www.npkua.org/pdf/PKU%20Binder%202011-Ch3.pdf. Some programs extend the target range to 10 mg/dl (600 μmol/L) for adults. See New England Consortium of Metabolic Programs, "PKU Toolkit," newenglandconsortium.org/toolkit/pku-issues.html.

6. For more information on these and other quoted comments from interviews with adolescents and adults with PKU, see our Note on Sources.

7. C. Bilginsoy, N. Waitzman, C. O. Leonard, and S. L. Ernst, "Living with Phenylketonuria: Perspectives of Patients and Their Families," *Journal of Inherited Metabolic Disease* 28, no. 5 (2005): 639–49, on 643.

8. Cary O. Harding, "'Mommy, Why Can't I Have a Hamburger Like the Other Kids?'" *Gene Therapy* 7, no. 23 (2000): 1969–70, on 1969.

9. Although much of the requirement is for the synthesis of new proteins, phenylalanine is also involved in neurotransmitter, catecholamine, and melanin synthesis. Exactly how much of the phenylalanine we ingest is excess will vary depending on individuals' specific phenylalanine requirements and dietary habits, such as how much carbohydrate and fat they ingest and whether they are vegetarians or occasional carnivores or largely subsist on burgers and brats. Cary Harding, personal communication, Jan. 6, 2012.

10. Actual intake is about 70 grams for women and 102 grams for men. US Department of Agriculture, Agricultural Research Service, "Table 2: Nutrient Intakes from Food," National Health and Nutrition Examination Survey 2005–2006 (2008), www.ars.usda.gov/SP2UserFiles/Place/12355000/pdf/0506/Table_2_NIF_05.pdf. A typical American adult weighing 150 pounds needs about 2,250 mg of phenylalanine daily.

11. Donna A. Messner, "On the Scent: The Discovery of PKU," *Chemical Heritage*, spring 2012, 32–36, on 34.

12. The warning labels seem to be unique to the United States, although many countries require that phenylalanine be included in the list of a product's ingredients.

13. The American Cancer Society estimates that a can of diet soda contains about 180 mg of aspartame. American Cancer Society, "Aspartame," www.cancer.org/Cancer/CancerCauses/OtherCarcinogens/AtHome/aspartame. According to the Coca-Cola Company, an 8 fl. oz. serving of Diet Coke has 125 mg of aspartame, with 60 mg of phenylalanine. Coca-Cola Company Beverage Institute for Health and Wellness, "What Health Professionals Need to Know about Aspartame,"

beverageinstitute.org/us/expert/what-health-professionals-need-to-know-about-aspartame. (However, soda cans typically contain 12 fl. oz.) In Canada and other countries where aspartame is combined with other synthetic sweeteners, the phenylalanine content is lower.

14. Anita MacDonald, "Diet and Compliance in Phenylketonuria," *European Journal of Pediatrics* 159, suppl. 2 (2000): S136–41, on S139.

15. Kristen Ahring, Amaya Bélanger-Quintana, Katharina Dokoupil, et al., "Dietary Management Practices in Phenylketonuria across European Centres," *Clinical Nutrition* 28, no. 3 (2009): 231–36.

16. Virginia E. Schuett, *Low Protein Food List for PKU*, 3rd ed. (CreateSpace, 2010); "free foods" are listed on pp. 29–37.

17. Ibid.; fruits and vegetables are listed on pp. 47–66.

18. Christine M. Trahms Program for Phenylketonuria, Seattle, "What Is the Diet for PKU?" depts.washington.edu/pku/about/diet.html. Botanically, potatoes are vegetables, but as they substitute for other starchy carbohydrates such as rice and pasta, they are often categorized as starches.

19. Mead Johnson Metabolics, "Adult PKU: Important Information for You," www.mjn.com/professional/pdf/LF873REV-10-05.pdf.

20. Texas Department of Health, *A Teacher's Guide to PKU*, developed by Mimi Kaufman and Maria Nardella (1985), www.pkuil.org/TeachersGuide.pdf.

21. Neil R. M. Buist, Kathleen Huntington, and Susan C. Winter, "Healthcare Coverage for Medical Food Treatment for Inborn Errors of Metabolism," NORD (National Organization for Rare Disorders (June 23, 2009), 9, www.rarediseases.org/docs/medical-foods/buist-winters-huntington–2011/view.

22. Louis I. Woolf, Remarks at 8th PKUCS Conference, Feb. 28–29, 1972, 199–200, Guthrie papers, box 15.

23. Frances J. Rohr, personal communication, Sept. 13, 2004.

24. Internationally, there is no agreement on the optimal dosage. A. MacDonald, A. Daly, P. Davies, et al., "Protein Substitutes for PKU: What's New?" *Journal of Inherited Metabolic Disease* 27 (2004): 363–71, on 364–65.

25. Neil R. M. Buist, interview, Portland, OR, Sept. 4, 2007.

26. Sandra C. Van Calcar, Erin L. MacLeod, Sally T. Gleason, et al., "Improved Nutritional Management of Phenylketonuria by Using a Diet Containing Glycomacropeptide Compared with Amino Acids," *American Journal of Clinical Nutrition* 89, no. 4 (2009): 1068–77.

27. Anita MacDonald, "Diet and Compliance in Phenylketonuria," *European Journal of Pediatrics* 159, suppl. 2 (2000): S136–41, on S140; Susan Waisbren, personal communication, July 27, 2012.

28. Francois Feillet, Anita MacDonald, Danielle Hartung (Perron), and Barbara Burton, "Outcomes beyond Phenylalanine: An International Perspective," *Molecular Genetics and Metabolism* 99, suppl. 1 (2010): S79–85, on S79.

29. Amber E. ten Hoedt, Heleen Maurice-Stam, Carolien C. A. Boelen, et al.,

"Parenting a Child with Phenylketonuria or Galactosemia: Implications for Health-Related Quality of Life," *Journal of Inherited Metabolic Disease* 34, no. 2 (2011): 391–98.

30. A. MacDonald, G. Harris, G. Rylance, et al., "Abnormal Feeding Behaviours in Phenylketonuria," *Journal of Human Nutrition and Dietetics* 10, no. 3 (1997): 163–70, on 169.

31. "Interviewees" refers collectively to current or former patients of the metabolic clinic at Boston Children's Hospital and to adults with PKU who were more informally interviewed by DBP at a "StoryCorps" booth at the International Conference for Adults and Teens with PKU, Chicago, Aug. 15–16, 2008.

32. MacDonald, "Diet and Compliance," S140.

33. Of 160 European centers surveyed, 65 percent included a psychologist, compared with fewer than half of centers surveyed in the United States. (The exact figure was not reported.) Feillet et al., "Outcomes beyond Phenylalanine," S80.

34. J. H. Walter, F. J. White, S. K. Hall, et al., "How Practical Are Recommendations for Dietary Control in Phenylketonuria?" *Lancet* 369 (2002): 55–57, on 56; Anita MacDonald, Hulya Gokmen-Ozel, Margareet van Rijn, and Peter Burgard, "The Reality of Dietary Compliance in the Management of Phenylketonuria," *Journal of Inherited Metabolic Disease* 33 (2010): 665–70, on 667.

35. MacDonald et al., "Reality of Dietary Compliance," 665. Although the term "compliance" remains widely used in the medical literature, it has been criticized as paternalistic, and "adherence" or, in the United Kingdom, "concordance" is increasingly taking its place. We generally use "adherence," except when citing the work of others, where we retain the original terminology.

36. Ibid.

37. Eduardo Sabaté, ed., *Adherence to Long-Term Therapies: Evidence for Action* (Geneva: World Health Organization, 2003), www.who.int/chp/knowledge/publications/adherence_report/en.

38. Adherence generally is inversely proportional to frequency of dose. See A. J. Claxton, J. Cramer, and C. Pierce, "A Systematic Review of the Associations between Dose Regimens and Medication Compliance," *Clinical Therapeutics* 23 (2001): 1296–310; Lars Osterberg and Terrence Blaschke, "Adherence to Medication," *New England Journal of Medicine* 353, no. 3 (2005): 487–97; Sameer D. Saini, Philip Schoenfeld, Kellee Kaulback, and Marla C. Dubinsky, "Effect of Medication Dosing Frequency on Adherence in Chronic Diseases," *American Journal of Managed Care* 15, no. 6 (2009): e22–33.

39. This applies to both type 1 diabetes (insulin-dependent, childhood-onset, requiring insulin treatment) and type 2 diabetes (non-insulin-dependent, adult-onset, manageable with diet alone). Jane Ogden, *The Psychology of Eating: From Healthy to Disordered Behavior*, 2nd ed. (Hoboken, NJ: Wiley-Blackwell, 2010), 19. For statistics on adherence in diabetes, see Sabaté, *Adherence to Long-Term Therapies*, 95–96.

40. Susan E. Waisbren, "Phenylketonuria," in *Handbook of Neurodevelopmental and Genetic Disorders in Children*, 2nd ed., ed. Sam Goldstein and Cecil R. Reynolds (New York: Guildford Press, 2011), 398–424, on 402. The impact of reduced blood phenylalanine levels on cognitive functioning is difficult to assess, as adults returning to the diet will also typically be returning to the clinic, sending in blood samples, receiving help with insurance, and so forth. The evidence on executive function is mixed; see M. L. Lindegren, S. Krishnaswami, C. Fonnesbeck, et al., *Adjuvant Treatment for Phenylketonuria (PKU)*, Comparative Effectiveness Review No. 56, prepared by the Vanderbilt Evidence-Based Practice Center under Contract No. HHSA 290-2007-10065-I, AHRQ Publication No. 12-EHC035-EF (Rockville, MD: Agency for Healthcare Research and Quality, Feb. 2012), www.effectivehealthcare.ahrq.gov/search-for-guides-reviews-and-reports/?pageaction=displayproduct&productid=957. On the impact on processing speed, see Julia Albrecht, Sven F. Garbade, and Peter Burgard, "Neuropsychological Speed Tests and Blood Phenylalanine Levels in Patients with Phenylketonuria: A Meta-Analysis," *Neuroscience and Biobehavioral Reviews* 33, no. 3 (2009): 414–22. On the reversibility of demyelination (associated with PKU) with a strict low-phe diet, see Peter J. Anderson and Vincenzo Leuzzi, "White Matter Pathology in Phenylketonuria," *Molecular Genetics and Metabolism* 99, suppl. (2010): S3–9.

41. Feillet et al., "Outcomes beyond Phenylalanine," S82.

42. G. M. Enns, R. Koch, V. Brumm, et al., "Suboptimal Outcomes in Patients with PKU Treated Early with Diet Alone: Revisiting the Evidence," *Molecular Genetics and Metabolism* 101 (2010): 99–109 (a meta-review of 150 outcome studies published since 2000). See also Waisbren, "Phenylketonuria," 404–9.

43. Rosa Gassió, Raphael Artuch, Maria Antonia Vilaseca, et al., "Cognitive Functions and the Antioxidant System in Phenylketonuric Patients," *Neuropsychology* 22, no. 4 (2008): 426–31.

44. Rosa Gassió, Eugenia Fusté, Anna López-Sala, et al., "School Performance in Early and Continuously Treated Phenylketonuria," *Pediatric Neurology* 33, no. 4 (2005): 267–71. For a review of studies on school achievement, see Waisbren, "Phenylketonuria."

45. Waisbren, "Phenylketonuria," 413.

46. Susan Waisbren notes that there may be a threshold effect, so fluctuations may not matter for individuals whose blood phenylalanine levels are well above the target range. Personal communication, June 27, 2012.

47. Cary O. Harding, personal communication, Jan. 6, 2012. See also Cary Harding, "Recent Advances in Cell and Gene Therapy for PKU," *National PKU News* 16, no. 2 (2004): 1–5.

48. Vera Anastasoaie, Laura Kurzius, Peter Forbes, and Susan Waisbren, "Stability of Blood Phenylalanine Levels and IQ in Children with Phenylketonuria," *Molecular Genetics and Metabolism* 95 (2008): 17–20.

49. Waisbren, "Phenylketonuria," 416.

50. F. J. van Spronsen and P. Burgard, "The Truth of Treating Children with Phenylketonuria after Childhood: The Need for a New Guideline," *Journal of Inherited Metabolic Disease* 31 (2008): 673–79, on 674.

51. Enns et al., "Suboptimal Outcomes"; Waisbren, "Phenylketonuria," 408–9 (citing results of the German Collaborative Study of PKU); M. P. A. Hoeks, M. den Heijer, and M. C. H. Janssen, "Adult Issues in Phenylketonuria," *Netherlands Journal of Medicine* 67, no. 1 (2009): 2–7. The etiology of bone disease in PKU is unclear, with some studies finding a correlation between low bone mineral density and dietary control and others not. See Enns et al., "Suboptimal Outcomes," 103; H. M. Koura, Ismail N. Abdallah, A. F. Kamel, et al., "A Long-Term Study of Bone Mineral Density in Patients with Phenylketonuria under Diet Therapy," *Archives of Medical Science* 7, no. 3 (2011): 493–500; A. B. Mendes, F. F. Martins, W. M Cruz, et al., "Bone Development in Children and Adolescents with PKU," *Journal of Inherited Metabolic Disease* 35, no. 3 (2012): 425–30.

52. The astrophysicist's website, www.astro.sunysb.edu/tracy/mystory.html, mentions the pediatrician. The mathematician maintains a website at engineering.purdue.edu/~mboutin/pku.html.

53. Waisbren, "Phenylketonuria," 416.

54. Buist interview.

55. van Spronsen and Burgard, "Truth of Treating Patients," 676.

56. Margaret Visser, *Much Depends on Dinner* (New York: Collier, 1986).

57. David A. Booth, *Psychology of Nutrition* (Bristol, PA: Taylor and Francis, 1994), 83.

58. Mary Douglas, "Deciphering a Meal," in *Implicit Meanings: Essays in Anthropology*, 2nd ed. (London: Routledge, 1999), 249–75.

59. Singh quoted in James Angelos, "The Great Divide," *New York Times*, Feb. 22, 2009.

60. Mary Douglas, "Standard Social Uses of Food," in *Food in the Social Order: Studies of Food and Festivities in Three American Communities*, ed. Mary Douglas (New York: Russell Sage Foundation, 1984), 1–39, on 10.

61. E. Vegni, L. Fiori, E. Riva, et al., "How Individuals with Phenylketonuria Experience Their Illness: An Age-Related Qualitative Study," *Child Care, Health and Development* 36 (2010): 539–48.

62. Frances J. Rohr, interview, Boston, Mar. 7, 2007.

63. Eating Disorders Review.com, "Fast Food: Now About One-Third of Teens' Meals Away from Home," first published in *Eating Disorders Review* 13 (Mar./Apr. 2003), www.eatingdisordersreview.com/nl/nl_edr_13_2_6.html.

64. Schuett, *Low Protein Food List*, "Convenience Foods," 219–44.

65. Annemarie Mol, *The Logic of Care: Health and the Problem of Patient Choice* (London: Routledge, 2008), 11, 23.

66. Mead Johnson Metabolics, "Adult PKU."

67. van Spronsen and Burgard, "Truth of Treating Patients." The authors note that although testing for PKU dates to 1963, in many countries, the program was im-

plemented years later and "no one knows what will really happen with early treated patients after the age of 40 years" (674). See also Nicole Frank, Ruth Fitzgerald, and Michael Legge, "Phenylketonuria—The Lived Experience," *New Zealand Medical Journal* 120, no. 1262 (2007): 2728, journal.nzma.org.nz/journal/120-1262/2728.

68. For an overview of this history, see Rani H. Singh, "The Enigma of Medical Foods," *Molecular Genetics and Metabolism* 92 (2007): 3–5.

69. Camp et al., "Nutritional Treatment," 4.

70. Buist et al., "Healthcare Coverage," table 2, p. 10.

71. The Cambrooke Foods website includes an updated list. Cambrooke Foods, "State Legislation," www.cambrookefoods.com/let_us_help/state_legislation.php.

72. Ibid., 7.

73. Cambrooke Foods, "State Legislation," 5.

74. Bilginsoy et al., "Living with Phenylketonuria," 643–44.

75. Phyllis B. Acosta, interview, Atlanta, Dec. 7, 2007.

76. Rohr interview.

77. European Society for Phenylketonuria and Allied Disorders Treated as Phenylketonuria (ESPKU), "Reimbursement PKU Products in Europe" (Mar. 28, 2011), www.espku.org/images/stories/reimbursement_ESPKU_2011_vs3.pdf.

78. Mireille Boutin, "Mimi's PKU Page," engineering.purdue.edu/~mboutin/pku.html. Boutin directs the Computational Imaging Laboratory at Purdue. For Boutin's biography, see engineering.purdue.edu/~mboutin/bio.html. Parents are taught to follow such practices early in their education about the management of their child's disorder, but Boutin would be in a minority who continue to hew so closely to the recommendations as adults.

Chapter 8. The Perplexing Problem of Maternal PKU

1. Editorial, "Maternal Phenylketonuria," *New England Journal of Medicine* 275 (Dec. 15, 1966): 1379–80.

2. Through the 1950s, some females with PKU were sterilized, with the support of some metabolic centers, probably at the request of parents who feared having to care for a mentally retarded daughter and grandchild. Phyllis Acosta, interview, Emory University, Atlanta, Dec. 7, 2007. A 1983 study of the effects of untreated maternal PKU and mild hyperphenylalaninemia on the fetus provides evidence that few women with severe forms of the disease reproduced. To avoid clinical bias of ascertainment in their study, the investigators identified women with maternal elevated blood phenylalanine levels by screening umbilical cord blood specimens obtained during deliveries in Massachusetts. Of the 22 women thus identified, only 2 had "classical" PKU. The small number indicates that, prior to screening, women with severe forms of the disease were usually severely impaired and only rarely bore children. Harvey L. Levy and Susan E. Waisbren, "Effects of Untreated Maternal Phenylketonuria and Hyperphenylalaninemia on the Fetus," *New England Journal of Medicine* 309, no. 21 (1983): 1269–74.

3. William B. Hanley, "Finding the Fertile Woman with Phenylketonuria," *European Journal of Obstetrics and Gynecology and Reproductive Biology* 137 (2008): 131–35, on 134.

4. "ACOG Committee Opinion, Number 449: Maternal Phenylketonuria," *Obstetrics and Gynecology* 114, no. 6 (2009): 1432–33, on 1432.

5. Increasingly, the condition with high phenylalanine levels during pregnancy, or maternal PKU (MPKU), is distinguished from the effects on the fetus, now called "maternal PKU syndrome" (MPKUS).

6. Asbjørn Følling, "The Original Detection of Phenylketonuria," in *Phenylketonuria and Some Other Inborn Errors of Amino Acid Metabolism*, ed. Horst Bickel, Fred P. Hudson, and Louis I. Woolf (Stuttgart: Georg Thieme, 1971), 1–3, on 3.

7. C. E. Dent, "Discussion of Paper by Armstrong, M.D.: Relation of Biochemical Abnormality to Development of Mental Defect in Phenylketonuria," in *Etiologic Factors in Mental Retardation: A Symposium Held at Chapel Hill, North Carolina, Nov. 8–9, 1956. Report of the 23rd Ross Pediatric Research Conference* (Columbus, OH: Ross Laboratories, 1957).

8. J. C. Denniston, "Children of Mothers with Phenylketonuria," *Journal of Pediatrics* 63 (1963): 461–62; C. C. Mabry, J. C. Denniston, and J. G. Coldwell, "Maternal Phenylketonuria: Mental Retardation in Children without Metabolic Defect," *New England Journal of Medicine* 269 (1963): 1404–8.

9. Robert Guthrie to All Maternal and Child Health Directors and Laboratory Directors Who Have Participated in the Trial of the "Inhibition Assay" for Early Detection of Phenylketonuria, Jan. 8, 1964, USCB, NCEMCH, box 071653.

10. C. C. Mabry, J. C. Denniston, and J. G. Coldwell, "Mental Retardation in Children of Phenylketonuric Mothers," *New England Journal of Medicine* 275 (1966): 1331–36.

11. See, for example, B. W. Richards, "Maternal Phenylketonuria," *Lancet* 1 (Apr. 11, 1964): 829; R. W. Coffelt, "Unexpected Finding from a PKU Newborn Screening Program," *Pediatrics* 34 (1964): 889–90; R. O. Fisch, W. A. Walker, and J. A. Anderson, "Prenatal and Postnatal Developmental Consequences of Maternal Phenylketonuria," *Pediatrics* 37 (1966): 979–86; R. E. Stevenson and C. C. Huntley, "Congenital Malformations in Offspring of Phenylketonuric Mothers," *Pediatrics* 40 (1967): 33–45; W. K. Frankenburg, B. R. Duncan, R. W. Coffelt, et al., "Maternal Phenylketonuria: Implications for Growth and Development," *Journal of Pediatrics* 73 (1968): 560–70; C. C. Huntley and R. E. Stevenson, "Maternal Phenylketonuria: Course of Two Pregnancies," *Obstetrics and Gynecology* 34 (1969): 694–700; R. O. Fisch, D. Doeden, L. L. Lansky, and J. A. Anderson, "Maternal Phenylketonuria: Detrimental Effects on Embryogenesis and Fetal Development," *American Journal of Diseases of Children* 118 (1969): 847–58; J. S. Yu and M. T. O'Halloran, "Children of Mothers with Phenylketonuria," *Lancet* 1 (1970): 210–12; H. Hansen, "Epidemiological Considerations on Maternal Hyperphenylalaninemia," *American Journal of Mental Deficiency* 75 (1970): 22–25; H. L. Levy, V. Karolkewicz, S. A. Houghton,

and R. A. MacCready, "Screening the 'Normal' Population in Massachusetts for Phenylketonuria," *New England Journal of Medicine* 282 (1970): 1455–58. Many of these reports are summarized in R. A. MacCready and H. L. Levy, "The Problem of Maternal Phenylketonuria," *American Journal of Obstetrics and Gynecology* 113 (1972): 121–28, which also reports on additional cases.

12. See, for example, D. E. Boggs and H. A. Waisman, "Influence of Excess Dietary Phenylalanine on Pregnant Rats and Fetuses," *Proceedings of the Society for Experimental Biology and Medicine* 115 (1964): 407–10; G. R. Kerr, A. S. Chamove, H. F. Harlow, and H. A. Waisman, "Fetal PKU: The Effect of Maternal Hyperphenylalaninemia during Pregnancy in the Rhesus Monkey (*Macaca mulatta*)," *Pediatrics* 42, no. 1 (1968): 27–36.

13. For example, Robert Fisch and colleagues, noting that early detection and dietary treatment meant that more homozygous females would be bearing children, urged that greater attention "be paid not only to the detection and treatment of phenylketonuria but also to the management of phenylketonuric gravidas." Fisch et al., "Prenatal and Postnatal Developmental Consequences," 985.

14. Følling, "Original Detection," 1–3; Holger Hansen, "Risk of Fetal Damage in Maternal Phenylketonuria," *Journal of Pediatrics* 83 (1973): 506–7.

15. Editorial, "Maternal Phenylketonuria," *Lancet*, Mar. 14, 1964, 598. The editorial was prompted by the original study by Mabry and colleagues. Mabry et al., "Maternal Phenylketonuria."

16. Editorial, "Maternal Phenylketonuria" (1966).

17. See, for example, Fisch et al., "Prenatal and Postnatal Developmental Consequences," 979–86; N. P. Forbes, K. N. F. Shaw, R. Koch, et al., "Maternal Phenylketonuria," *Nursing Outlook* 14 (1966): 40–42; Mabry et al., "Mental Retardation," 1331–36; MacCready and Levy, "Problem of Maternal Phenylketonuria," 121–28.

18. Neil R. M. Buist, interview, Portland, OR, Sept. 4, 2007.

19. Acosta interview.

20. Stevenson and Huntley, "Congenital Malformations," 44.

21. In that case, a woman was maintained on a low-phe diet during the final five months of her fourth pregnancy; although the infant's measurements were below normal, he was developmentally more advanced than his siblings at comparable ages. J. D. Allan and J. K. Brown, "Maternal Phenylketonuria and Foetal Brain Damage: An Attempt at Prevention by Dietary Control," in *Some Recent Advances in Inborn Errors of Metabolism*, ed. K. S. Holt and V. P. Coffey (London: Livingston, 1968), 14.

22. Huntley and Stevenson, "Maternal Phenylketonuria," 699.

23. I. Smith, M. Erdohazi, F. J. Macartney, et al., "Fetal Damage Despite Low-Phenylalanine Diet after Conception in a Phenylketonuric Woman," *Lancet* 1 (1979): 17–19; K. B. Nielsen, E. Wamberg, and J. Weber, "Successful Outcome of Pregnancy in a Phenylketonuric Woman after Low-Phenylalanine Diet Introduced before Conception," *Lancet* 1 (1979): 1245.

24. Hansen, "Epidemiological Considerations." In an interview, Harvey Levy commented that he was fascinated at the time by Hansen's idea that mental retardation in the baby is not determined by the mother's biochemical defect, but he noted that "nobody since then has been able to look at a large number of women who've gone through maternal PKU pregnancies untreated, which is what you'd have to have. Once you get treatment into the mix, then all bets are off." Harvey L. Levy, interview, Newton, MA, Dec. 8, 2006.

25. H. A Waisman, "Role of Hyperphenylalaninemia in Pregnant Women as a Cause of Mental Retardation in Offspring," *American College of Obstetrics and Gynecology* 99 (1967): 431–33. See also N. R. M. Buist and B. M. Jhaveri, "A Guide to Screening Newborn Infants for Inborn Errors of Metabolism," *Journal of Pediatrics* 82, no. 3 (1973): 511–22.

26. S. P. Bessman, R. A. Wapnir, H. S. Pankratz, and B. A. Plantholt, "Maternal Phenylalanine Deprivation in the Rat: Enzymatic and Cellular Liver Changes in the Offspring," *Biologia Neonatorum* 14 (1969): 107–66, on 107.

27. Fisch et al., "Prenatal and Postnatal Developmental Consequences," 979; T. A. Munro, "Phenylketonuria: Data on Forty-Seven British Families," *Annals of Eugenics* 14 (1947): 60–68.

28. Harvey L. Levy, "Historical Background for the Maternal PKU Syndrome," *Pediatrics* 112 (2003): 1516–18, on 1516.

29. For example, "The implications for girls and pregnancy may influence the decision regarding the time to discontinue the diet." Richard Koch, Elizabeth Wenz, and Mary Smartt Steinberg, *PKU: A Guide to Management* (Sacramento: Bureau of Maternal and Child Health, California Department of Health, 1977), 21.

30. Mary Egan to Ellen S. Kang, Jan. 4, 1967, USCB, NARA II, box 1000, file 4-5-11-5.

31. T. L. Perry, S. Hansen, B. Tischler, et al., "Unrecognized Adult Phenylketonuria—Implications for Obstetrics and Psychiatry," *New England Journal of Medicine* 289 (1973): 395–98. Thomas Perry was an especially strong advocate for the position that women with high blood phenylalanine levels should be counseled, in the first instance, to abort. Levy interview. Perry's position recommending therapeutic abortion was critiqued in a letter to the editor by Levy and defended in a letter by Charles Johnson of the University of Iowa's Child Development Clinic. Both letters and Perry's reply appear in "Management of Women with Phenylketonuria," *New England Journal of Medicine* 290, no. 2 (1974): 108–9.

32. H. Hansen, "Risk of Fetal Damage in Maternal Phenylketonuria," *Journal of Pediatrics* 83 (1973): 506. Robert Fisch similarly asserted that "the presently available dietary treatment of pregnant phenylketonuric women is of questionable value, and phenylketonuric women should possibly be advised not to have children." Fisch et al., "Maternal Phenylketonuria," 855.

33. J. M. G. Wilson and F. Jungner, *Principles and Practice of Screening for Disease* (Geneva: World Health Organization, 1968), 29.

34. For example, Buist and Jhaveri argued that a policy that promoted pregnancy termination would potentially abort as many normal fetuses as those severely affected and that a better approach would be some form of phe-restricted diet. N. R. M. Buist and B. M. Jhaveri, Letter to the Editor, *Journal of Pediatrics* 83 (1973): 507.

35. Susan E. Waisbren, Children's Hospital Boston, interview, Boston, Feb. 8, 2007.

36. Frances J. Rohr, Children's Hospital Boston, interview, Boston, Mar. 7, 2007.

37. K. S. Tice, E. Wenz, and K. Jew, "Reproductive Counseling for Adolescent Females with Phenylketonuria," *Journal of Inherited Metabolic Disease* 3 (1980): 105–7.

38. R. R. Lenke and H. L. Levy, "Maternal Phenylketonuria and Hyperphenylalaninemia: An International Survey of the Outcome of Untreated and Treated Pregnancies," *New England Journal of Medicine* 303 (1980): 1202–8.

39. Levy and Waisbren, "Effects of Untreated Maternal Phenylketonuria."

40. H. N. Kirkman and D. M. Frazier, "Maternal PKU: Thirteen Years after Epidemiological Projections," *International Pediatrics* 11, no. 5 (1996): 279–83. More recent data suggest that the degree, although not incidence, of mental retardation is less than in PKU, probably because in PKU there is continuing exposure to the toxic substance, whereas in MPKU the exposure lasts only nine months. Prenatal exposure to phenylalanine appears to be associated with high levels of hyperactivity and other behavioral problems, and postnatal exposure with greater difficulty in metacognitive tasks and slower processing speed. K. M. Antshel and S. E. Waisbren, "Timing Is Everything: Executive Functions in Children Exposed to Elevated Levels of Phenylalanine," *Neuropsychology* 17, no. 3 (2003): 458–68; K. M. Antshel and S. E. Waisbren, "Developmental Timing of Exposure to Elevated Levels of Phenylalanine Is Associated with ADHD Symptom Expression," *Journal of Abnormal Child Psychology* 31, no. 6 (2003): 565–74.

41. H. N. Kirkman, "Projections of a Rebound in Frequency of Mental Retardation from Phenylketonuria," *Applied Research in Mental Retardation* 3 (1982): 319–28. Kirkman had published briefly on the issue in 1979, but this one-paragraph précis of a paper presented at a meeting of the Society for Pediatric Research had little if any impact. H. N. Kirkman, "Projections of Mental Retardation from PKU," *Pediatric Research* 13 (1979): 414.

42. Lionel S. Penrose, "Phenylketonuria: A Problem in Eugenics," *Lancet* 1 (1946): 949–53; Kirkman, "Projections of a Rebound."

43. The calculations assume a population of fixed numbers over time. Kirkman, "Projections of a Rebound," 322–23; Henry N. Kirkman, interview, Chapel Hill, NC, Feb. 12, 2006.

44. "Gravidity" refers to the number of times a woman has been pregnant. Kirkman, "Projections of a Rebound."

45. S. L. C. Woo, A. S. Lidsky, F. Güttler, et al., "Cloned Human Phenylalanine Hydroxylase Gene Allows Prenatal Diagnosis and Carrier Detection of Classical Phenylketonuria," *Nature* 306 (Nov. 10, 1983): 151–55.

46. Felix de la Cruz, interview, Bethesda, MD, July 3, 2002. At the time, de la Cruz was chief of the Mental Retardation and Developmental Disabilities Branch, NICHD.

47. Due to ascertainment bias—most women with hyperphenylalaninemia would not have been treated and therefore would have been lost to follow-up—there was uncertainty as to whether their offspring were also at risk. However, a study that avoided such bias by using umbilical cord blood specimens found that mild hyperphenylalaninemia in the mother had few cognitive or other effects on the fetus. H. L. Levy and S. E. Waisbren, "Effects of Untreated Maternal Phenylketonuria and Hyperphenylalaninemia on the Fetus," *New England Journal of Medicine* 309 (1983): 1269–74.

48. In this period, 80 to 90 percent of women who were never treated—nearly all of whom were unmarried and unemployed—aborted their pregnancies. According to an MPKUCS report, of 358 patients for whom treatment was prescribed, 48 were never treated. This group included 6 spontaneous abortions and 38 elective terminations. Depending on whether or not spontaneous abortions are subtracted from the total, the elective termination rate is 80 or 90 percent. R. Koch, H. L. Levy, R. Matalon, et al., "International Collaborative Study of Maternal Phenylketonuria: Status Report 1994," *Acta Paediatrica* 407, suppl. (1994): 111–19, on 113–14. See also R. Koch, H. L. Levy, R. Matalon, et al., "The North American Collaborative Study of Maternal Phenylketonuria: Status Report 1993," *American Journal of Diseases of Childhood* 147 (1993): 1224–30, on 1226.

49. In the United States, the recommended range of blood phenylalanine during pregnancy is 120 to 360 µmol/L (2–6 mg/dL under the older method of calculation).

50. Patient interview, June 19, 2007.

51. Patient interview, Mar. 22, 2007.

52. A. S. Brown, P. M. Fernhoff, S. E. Waisbren, et al., "Barriers to Successful Dietary Control among Pregnant Women with Phenylketonuria," *Genetics in Medicine* 4 (2002): 84–89. It is notable that the most recent US studies are from 2002.

53. Kirkman and Frazier, "Maternal PKU," 279–83. They note that only 9 percent of mothers who were being treated elected to terminate their pregnancy, in contrast to 90 percent of untreated mothers (282). The rebound issue has recently been reanalyzed in Robert Resta, "Generation n + 1: Projected Numbers of Babies Born to Women with PKU Compared to Babies with PKU in the United States in 2009," *American Journal of Medical Genetics A* 158A, no. 5 (2012): 1118–23.

54. A study by the Centers for Disease Control and Prevention, based on interviews with 30 women, found that 71 percent discontinued the diet after pregnancy. Brown et al., "Barriers to Successful Dietary Control," 87. Another study found

that 91 percent of women discontinued the diet. F. Rohr, A. Munier, D. Sullivan, et al., "The Resource Mothers Study of Maternal Phenylketonuria: Preliminary Findings," *Journal of Inherited Metabolic Disease* 27 (2004): 145–55.

55. Patient interview, June 21, 2006.

56. R. Koch, F. Trefz, and S. Waisbren, "Psychosocial Issues and Outcomes in Maternal PKU," *Molecular Genetics and Metabolism* 99, suppl. 1 (2010): S68–74.

57. T.-W. Ng, A. Rae, H. Wright, et al., "Maternal Phenylketonuria in Western Australia: Pregnancy Outcomes and Developmental Outcomes in Offspring," *Journal of Paediatrics and Child Health* 39, no. 5 (2003): 358–63, on 363.

Chapter 9. Who Should Procreate?

1. S. L. C. Woo, A. S. Lidsky, F. Güttler, et al., "Cloned Human Phenylalanine Hydroxylase Gene Allows Prenatal Diagnosis and Carrier Detection of Classical Phenylketonuria," *Nature* 306 (Nov. 10, 1983): 151–55.

2. There were tensions within the NARC, but they reflected not disagreements over segregation and sterilization but the organization's emphasis on a scientific approach to prevention, with some activists and officials believing that its focus on biological causes of mental retardation had deflected attention from more important factors. As noted in chapter 5, on his 1963 retirement as executive director of the NARC, Gunnar Dybwad provocatively advised parents to devote at least as much energy to promoting civic action programs to combat socioeconomic causes of mental retardation as to persuading state health departments to adopt universal testing for PKU. The persistence of such tensions is reflected in a letter from G. Vern Beckett, chair of the California Association for the Retarded (CAR), to the then NARC president, protesting a change in title of the proposed National Task Force on Prevention to the more limited Prevention from Bio-Medical Causes. "We can not forever pay lip service to prevention and, at the same time, continue to support those activities which, if implemented, would have little or no impact on the incidence of mental retardation or handicapped births," argued Beckett. Although the NARC leadership acknowledged that "the incidence of mental retardation from bio-medical causes is insignificant," it failed to act accordingly. Guthrie, who had been copied on the letter, responded that for the NARC to be politically effective in getting results, it must "select very specific, non-controversial causes of mental retardation where the knowledge of prevention is generally accepted and the means are available." G. Vern Beckett, Chair, CAR Research and Prevention Committee, to James R. Wilson, Jr., NARC President, Dec. 28, 1977; Robert Guthrie to G. Vern Beckett, Jan. 5, 1978; both in Guthrie papers, box 7, folder "NARC Task Force—1/76–present."

3. On Guthrie's campaign to prevent childhood lead poisoning, see Jean Holt Koch, *Robert Guthrie: The PKU Story* (Pasadena, CA: Hope Publishing House, 1970), 81–91.

4. Alison Bashford, "Epilogue: Where Did Eugenics Go?" in *The Oxford Hand-*

book of the History of Eugenics, ed. Alison Bashford and Philippa Levine (Oxford: Oxford University Press, 2010), 539–58, on 539.

5. Troy Duster's *Backdoor to Eugenics* (London: Routledge, 1990) is seminal, but numerous other authors have since made similar arguments.

6. Dorothy Nelkin and M. Susan Lindee, *The DNA Mystique: The Gene as Cultural Icon* (New York: W. H. Freeman, 1995), 34–37. See also Daniel Kevles, *In the Name of Eugenics: Genetics and the Uses of Human Heredity* (Cambridge, MA: Harvard University Press, 1995), 251–68; Edmund Ramsden, "Confronting the Stigma of Eugenics: Genetics, Demography and the Problems of Population," *Social Studies of Science* 39, no. 6 (2009): 853–84.

7. H. J. Muller, "Our Load of Mutations," *American Journal of Human Genetics* 2, no. 2 (1950): 111–76. See also Diane B. Paul, "'Our Load of Mutations' Revisited," *Journal of the History of Biology* 20, no. 3 (1987): 321–35. Muller's arguments gained strength in the 1950s with increased atmospheric testing of nuclear weapons.

8. Tracy M. Sonneborn, ed., *Control of Human Heredity and Evolution* (New York: Macmillan, 1965), 124.

9. Biochemist Rollin Hotchkiss coined this expression in his article "Portents for a Genetic Engineering," *Journal of Heredity* 56 (1965): 197–202. See also Lily E. Kay, *The Molecular Vision of Life: Caltech, the Rockefeller Foundation, and the Rise of the New Biology* (New York: Oxford University Press, 1993), esp. 275–76; Kevles, *In the Name of Eugenics*, 258–68; Susan Wright, *Molecular Politics: Developing American and British Regulatory Policy for Genetic Engineering, 1972–1982* (Chicago: University of Chicago Press, 1994), 123–24.

10. Joshua Lederberg, "Biological Future of Man," in *Man and His Future: A CIBA Foundation Volume*, ed. Gordon Wolstenholme (Boston: Little Brown, 1963), 264.

11. Crick and Pirie quoted in "Eugenics and Genetics: Discussion," in Wolstenholme, *Man and His Future*, 275, 282, http://archive.org/stream/manhisfutureciba00wols/manhisfutureciba00wols_djvu.txt.

12. Charles R. Scriver, "Treatment in Medical Genetics," in *Proceedings of the Third International Congress of Human Genetics*, ed. J. F. Crow and J. V. Neel (Baltimore: Johns Hopkins University Press, 1967), 45–56, on 54.

13. Theodosius Dobzhansky, *Mankind Evolving: The Evolution of the Human Species* (New Haven, CT: Yale University Press, 1962), 333. See also Dobzhansky's "On Genetic Aspects of Human Evolution," in Crow and Neel, *Proceedings*, 361–65.

14. Ashley Montagu, *Human Heredity* (Cleveland: World Publishing, 1959), 305–6.

15. Sheldon Reed, *Parenthood and Heredity* (New York: John Wiley, 1964), 85.

16. Albert R. Jonsen, *The Birth of Bioethics* (New York: Oxford University Press, 1998), 34–51.

17. Joseph F. Fletcher, "Ethical Aspects of Genetic Controls: Designed Genetic Changes in Man," *New England Journal of Medicine* 285 (1971): 776–83, on 782.

18. Ibid.

19. Joseph F. Fletcher, "Knowledge, Risk, and the Right to Reproduce," in *Genetics and the Law II*, ed. A. Milunsky and G. J. Annas (New York: Plenum Press, 1980), 131, 134.

20. Paul Ramsey, *Fabricated Man: The Ethics of Genetic Control* (New Haven, CT: Yale University Press, 1970), 98–99.

21. Phillip Handler, ed., *Biology and the Future of Man* (New York: Oxford University Press, 1970), 897. The Committee on Science and Public Policy, charged with preparing the report, began work in 1966. Curt Stern chaired the panel that drafted the final chapter, also titled "Biology and the Future of Man," from which the quotation is taken.

22. Paul R. Ehrlich, *The Population Bomb* (New York: Ballantine, 1968), 123–24. Other proposals to lower birth rates ranged from encouragement of voluntary family planning to schemes involving direct social controls; for a summary of suggestions, see Bernard Berelson, "Beyond Family Planning," *Studies in Family Planning* 1, no. 38 (1969): 1–16. On the intellectual roots of *The Population Bomb*, see Pierre Desrochers and Christine Hofauer, "The Post War Intellectual Roots of the Population Bomb: Fairfield Osborn's 'Our Plundered Planet' and William Vogt's 'Road to Survival' in Retrospect," *Electronic Journal of Sustainable Development* 1, no. 3 (2009). For a recent analysis of the links between the population control and environmental movements, including discussions of the views of Ehrlich and of Garrett Hardin, see Thomas Robertson, *The Malthusian Moment: Global Population Growth and the Birth of American Environmentalism* (New Brunswick, NJ: Rutgers University Press, 2012).

23. Kenneth Boulding, *The Meaning of the Twentieth Century* (New York: Harper and Row, 1964), 135.

24. Garrett Hardin, "The Tragedy of the Commons," *Science* 162, no. 3859 (1968): 1243–48, on 1248.

25. Linda Gordon, *Women's Body, Woman's Right: Birth Control in America*, rev. ed. (New York: Penguin Books, 1990), 386–97; Ramsden, "Confronting the Stigma"; Edmund Ramsden, "Eugenics from the New Deal to the Great Society: Genetics, Demography, and Population Quality," *Studies in the History and Philosophy of the Biological and Biomedical Sciences* 39 (2008): 391–406; Matthew Connelly, *Fatal Misconception: The Struggle to Control World Population* (Cambridge, MA: Harvard University Press, 2008), esp. chap. 5; Robertson, *Malthusian Moment*.

26. Vance Packard, *The People Shapers* (Boston: Little Brown, 1977), 257.

27. Charles Darwin, *The Descent of Man and Selection in Relation to Sex*, 2nd ed., introduction by James Moore and Adrian Desmond (London: Penguin Books, 2004 [1879]), 159.

28. Jonsen, *Birth of Bioethics*, 14.

29. A. G. Knudson, Jr., *Genetics and Disease* (New York: McGraw-Hill, 1965), 132.

30. As we discussed in chapter 8, in 1982, Henry Kirkman argued that any dysgenic effect of treating genetic disorders such as PKU would be negligible in the next few generations.

31. Handler, *Biology and the Future of Man*, 909, 910. The other treatable genetic disorders mentioned were galactosemia, fulminating juvenile diabetes, and pyloric stenosis.

32. P. B. Medawar and J. S. Medawar, *Aristotle to Zoos: A Philosophical Dictionary of Biology* (Cambridge, MA: Harvard University Press, 1985), 87.

33. P. B. Medawar and J. S. Medawar, "Eugenics," in *The Life Sciences: Current Ideas of Biology* (New York: Harper and Row, 1977), 56–65, on 60. Writing principally about the case of PKU, Peter Medawar also asserted, "New medical solutions will be found for the inborn disorders long before our present regimen of real or attempted prevention and cure can create an insupportable genetic burden." Peter Medawar, "The Genetical Impact of Medicine," *Annals of Internal Medicine* 67, no. 3 (1967): 28–31, on 30.

34. Peter Medawar, "Genetics and the Medicine of the Future," *Journal of the Mt. Sinai Hospital New York* 36 (1969): 189–93, on 193. This sentiment was repeated as late as 1983, when Medawar wrote, "To many, many people—especially those who have appointed themselves guardians of civil liberty and the conscience of us all—the idea that anyone should not marry whom he or she chooses is seen as a gross invasion of personal liberty, a wanton withdrawal of natural rights. This is a most questionable argument, for who has conferred upon human beings the right to bring genetically crippled children into the world?" P. B. Medawar and J. S. Medawar, "Phenylketonuria," in *Aristotle to Zoos*, 211–14, on 213.

35. Linus Pauling, "Biochemical and Structural Chemical Factors in Relation to Mental Disease, Especially Mental Deficiency," Aug. 1, 1955, courtesy of Ava Helen and Linus Pauling Papers, Special Collections, Oregon State University, Corvallis, OR, box 11.077.17. See pp. 7–8 of the grant application "The Search for a Test for Phenylketonuric Heterozygotes." On efforts to develop the test, see box 11.073.10.

36. Linus Pauling, "Orthomolecular Psychiatry," *Science* 160, no. 3825 (1968): 265–71.

37. Pauling quoted in "Nobel Prize Scientist at University of Oregon Asks for End to Nuclear Tests," *Register-Guardian* (Eugene, OR), Jan. 21, 1959, 6A.

38. Emile Zuckerkandl and Linus Pauling, "Molecular Disease, Evolution, and Genetic Heterogeneity," in *Horizons in Biochemistry*, ed. M. Kasha and B. Pullman (New York: Academic Press, 1962), 189–225, on 221.

39. Ibid., 221–22.

40. Linus Pauling, "Medicine in a Rational Society," *Journal of the Mt. Sinai Hospital New York* 36 (1969): 194–99. The original tattooing proposal is found in Linus Pauling, "Reflections on the New Biology: Foreword," *UCLA Law Review* 15 (1968): 267–72.

41. Lucy Eisenberg, "Genetics and the Survival of the Unfit," *Harper's Magazine*, Feb. 1966, 53–58, on 53.

42. Kingsley Davis, "Population Policy: Will Current Programs Succeed?" *Science* 158, no. 3802 (1967): 730–39, on 737.

43. Ramsden, "Confronting the Stigma," 860.

44. Donald Fleming, "On Living in a Biological Revolution," *Atlantic Monthly* 223 (1969): 64–70, on 69.

45. Ramsey, *Fabricated Man*, 1–2; Joseph Dancis, "The Prenatal Detection of Hereditary Defects," in *Medical Genetics*, ed. V. A. McKusick and R. Claiborne (New York: HP Publishing, 1973), 247; Gene Bylinsky, "What Science Can Do about Hereditary Disease," *Fortune* 90, no. 3 (1974): 148–52, on 152.

46. Paul Rabinow used this phrase (in a 1998 talk) in reference to French intellectuals' disquiet with recent developments in science and technology.

47. The Americans with Disabilities Act was passed in 1990, and the Disability Discrimination Act, in the United Kingdom, in 1995. The latter was partly supplanted by the Disability and Equality Act 2010. Australia also enacted antidiscrimination legislation in 1992.

48. Erving Goffman, *Asylums: Essays on the Social Situation of Mental Patients and Other Inmates* (New York: Anchor, 1961).

49. For more on the history of normalization, see Raymond Lemay, "Normalization and Social Role Valorization," in *Encyclopedia of Disability and Rehabilitation*, ed. A. E. Dell Orto and R. P. Marinelli (New York: Simon and Schuster/Macmillan, 1995), 515–21.

50. Joseph P. Shapiro, *No Pity: People with Disabilities Forging a New Civil Rights Movement* (New York: Random House, 1993). See also Doris Fleischer and Frieda Zames, *The Disability Rights Movement: From Charity to Confrontation*, 2nd ed. (Philadelphia: Temple University Press, 2011). In the most extreme view of the social model, all disability is attributed to societal factors, such as physical barriers, lack of support, or discriminatory attitudes. In more moderate versions, it is assumed that "people are disabled by society and their bodies." The wording is Tom Shakespeare's. FiveBooks, "FiveBooks Interviews: Tom Shakespeare on Disability" (Feb. 12, 2010), fivebooks.com/interviews/tom-shakespeare-on-disability. See also Tom Shakespeare, *Disability Rights and Wrongs* (Abingdon, UK: Routledge, 2006).

51. For a range of viewpoints in relation to prenatal diagnosis, see Erik Parens and Adrienne Asch, eds., *Prenatal Testing and Disability Rights* (Washington, DC: Georgetown University Press, 2000).

52. For example, Dr. Marie Hilliard, chair of the Committee on Ethics and Public Policy of the National Catholic Partnership, recently commented on the results of a survey indicating that 90 percent of members of the American Congress of Obstetricians and Gynecologists believe abortion is justified for fetal anomalies that are fatal and 63 percent believe so for anomalies that are not fatal: "Thus, nearly two-thirds of the physicians responding, physicians entrusted with the care of mother and unborn child, embrace eugenics. The very persons who are the guardians of the health of the unborn baby, the mother and the only professionals upon

whom that baby can rely, have become its judge, jury, and executioner for conviction of the 'crime' of being less than perfect." Marie Hilliard, "The New Eugenics: Eliminating the 'Undesirable' by Prenatal Diagnosis" (Aug. 10, 2010), www.ncpd.org/node/1141. Similarly, another antiabortion activist, referring to the killing of handicapped babies in ancient Greece and Rome, writes that "bigotry against people with disabilities is its [eugenics'] deepest bias at all, and possibly its oldest." Mary Meehan, "Triumph of Eugenics in Prenatal Testing: Part I. How It Happened," *Human Life Review* 35, no. 3 (2009): 28–40, on 29.

53. Kathryn Moss, "The 'Baby Doe' Legislation: Its Rise and Fall," *Policy Studies Journal* 15, no. 4 (1987): 629–51.

54. William Meadow, Ann Dudley Goldblatt, and John Lantos, "Current Opinion in Pediatrics: Ethics and Law 2001," *Current Opinion in Pediatrics* 14, no. 2 (2002): 170–73.

55. Gordon, *Woman's Body, Woman's Right*, 400–406.

56. Christine Di Stefano, "Autonomy in the Light of Difference," in *Revisioning the Political: Feminist Reconstructions of Traditional Concepts in Western Political Theory*, ed. Nancy J. Hirschmann and Christine Di Stefano (Boulder, CO: Westview Press, 1996), 95–116, on 95.

57. I. H. Porter, "Evolution of Genetic Counseling in America," in *Genetic Counseling*, ed. H. A. Lubs and F. de la Cruz (New York: Raven Press, 1977), 23; J. R. Sorenson and A. J. Culbert, "Genetic Counselors and Counseling Orientation: Unexamined Topics in Evaluation," in Lubs and de la Cruz, *Genetic Counseling*. For a detailed history of genetic counseling in the United States, see Alexandra M. Stern, *Telling Genes: The Story of Genetic Counseling in America* (Baltimore: Johns Hopkins University Press, 2012).

58. Diane B. Paul, "From Eugenics to Medical Genetics," *Journal of Policy History* 9, no. 1 (1997): 96–116; Nathaniel Comfort, *The Science of Human Perfection: How Genes Became the Heart of American Medicine* (New Haven, CT: Yale University Press, 2012), 97–129; Stern, *Telling Genes*, 75–89.

59. Stern, *Telling Genes*, 123–25. Stern discusses the intertwining of autonomy and nondirectiveness on pp. 143–45.

60. Lori B. Andrews, Jane E. Fullarton, Neil A. Holtzman, and Arno G. Motulsky, eds., *Assessing Genetic Risks: Implications for Health and Social Policy* (Washington, DC: National Academy Press, 1994), 103.

61. Thomas L. Beauchamp and James F. Childress, *Principles of Biomedical Ethics* (New York: Oxford University Press, 1979).

62. Jonsen, *Birth of Bioethics*, 335; Paul R. Wolpe, "The Triumph of Autonomy in American Bioethics: A Sociological View," in *Bioethics and Society: Sociological Investigations of the Enterprise of Bioethics*, ed. R. DeVries and J. Subedi J (Englewood Cliffs, NJ: Prentice-Hall, 1998), 43.

63. Beauchamp and Childress, *Principles of Biomedical Ethics*, 56–59, as quoted in Jonsen, *Birth of Bioethics*, 335. A voluminous literature now critiques this version

of autonomy. See especially Carl E. Schneider, *The Practice of Autonomy: Patients, Doctors, and Medical Decisions* (New York: Oxford University Press, 1998); Onora O'Neill, *Autonomy and Trust in Bioethics* (Cambridge: Cambridge University Press, 2002); Bruce Jennings, "Good-bye to All That—Autonomy," *Journal of Clinical Ethics* 13 (2002): 67–71; Alfred I. Tauber, "Sick Autonomy," *Perspectives in Biology and Medicine* 46 (2003): 484–95; G. M. Stirrat and R. Gill, "Autonomy in Medical Ethics after O'Neill," *Journal of Medical Ethics* 31 (2005): 127–30.

64. Ruth Faden, ed., *The Human Radiation Experiments: Final Report of the President's Advisory Committee* (New York: Oxford University Press, 1996), 80, 97–103.

65. He went on to note that "this shallowest meaning of respect for autonomy, unfortunately, seemed the most readily grasped." Jonsen, *Birth of Bioethics*, 335.

66. For example, participants at the 1962 Ciba Foundation meeting "Man and His Future" included such scientific luminaries as Crick, Muller, Julian Huxley, Lederberg, Peter Medawar, Pirie, J. B. S. Haldane, Albert Szent-Györgyi, and Gregory Pincus. Jonsen notes that 5 of the 27 participants were Nobel Prize winners. Ibid., 15.

67. Dorothy C. Wertz, "Society and the Not-So-New Genetics: What Are We Afraid Of? Some Future Predictions from a Social Scientist," *Journal of Contemporary Health Law and Policy* 13 (1997): 299–346. The survey question was highly hypothetical as there were no other legal choices in the third trimester.

68. Ibid., 339. Statistics are from data distributed by Dorothy Wertz at the "Workshop on Eugenic Thought and Practice: A Reappraisal towards the End of the Twentieth Century" (Van Leer Institute, Jerusalem, May 26–29, 1997).

Chapter 10. Newborn Screening Expands

1. Fluorometry was also used in Quebec and North Carolina to detect PKU. Although it was more sensitive than Guthrie's bacterial inhibition assay, it resulted in a larger number of false-positive results. Paper and thin-layer chromatography modified for use in NBS (with both blood and urine) had a lower sensitivity and specificity than tests for specific metabolites or enzymes but could be used to visualize several compounds.

2. Edwin W. Naylor, interview, Isle of Pines, SC, Oct. 23, 2008.

3. The machine originally punched the four discs from four different blood spots. On first seeing the machine, Guthrie suggested to Phillips that the four discs be punched from a single spot, which would save the other spots for further uses. The discs had initially been a quarter-inch thick. Phillips modified the machine pattern to punch discs three-sixteenths-inch thick instead, and Guthrie modified the assays to accommodate these smaller discs. Robert Guthrie, "The Origin of Newborn Screening," *Screening* 1 (1992): 5–15, on 13.

4. Bradford L. Therrell, interview, Austin, TX, Nov. 24, 2008. Dr. Therrell worked for more than 28 years for the Texas Department of Health Bureau of Labo-

ratories and later became director of the National Newborn Screening and Genetics Resource Center.

5. Three examples from 1965 are typical: "PKU Screening and Treatment Need Second Look, Says AAP," in the *Pediatric Herald* (reporting on a policy statement by the AAP's Committee on the Handicapped Child); "Caution Is Urged on Forced Use of PKU Tests," in the *Medical Tribune* (concerning the two infant deaths in Massachusetts [see chapter 5] and quoting Mary Efron's objections to mandated screening); and "Value of PKU Tests Questioned," in the Baltimore *Evening Sun* (featuring a photo and sympathetic account of the views of Bessman, who is quoted as saying that it would be better to test all children for heart disease and diabetes than for PKU). These and other critiques are in Guthrie papers, box 1, folder "Press Clippings, 1958–1980."

6. "After Ten Years of PKU Testing, a Re-evaluation," *Medical World News*, Nov. 19, 1971, 43–44, on 44.

7. Tabitha M. Powledge, "Genetic Screening as a Political and Social Development," in *Ethical, Social, and Legal Dimensions of Screening for Human Genetic Disease*, ed. Daniel Bergsma (New York: Stratton, 1974), 25–55, on 31. On Washington, DC, see memo from Polly Shackleton, Chairperson, Committee on Human Resources, to All Councilmembers, "Report on Bill #3-126, the District of Columbia Newborn Screening Requirement Act of 1979," Jan. 2, 1980, p. 2, Guthrie papers, box 9, folder "Washington, D.C." Since the incidence in the African American population was thought to be very low, the cost-effectiveness of screening would presumably improve if nonwhite babies were excluded; indeed, they had been omitted from the USCB field trials for this reason. However, a race-based approach to screening was ultimately rejected. For a discussion of the issue, including the controversy it engendered within the Children's Bureau, see M. Susan Lindee, *Moments of Truth in Genetic Medicine* (Baltimore: Johns Hopkins University Press, 2005), 37–39.

8. Memo, Robert Guthrie to Jerome A. Lackner, Director of Health, State of California, May 26, 1976, Guthrie papers, box 37, folder "Letters (photocopies) re: Legislation 1973, 1976." Guthrie charged that George Cunningham wrongly considered each test separately in terms of incidence and cost-effectiveness, when the major cost—that of collecting and transporting the specimen—had already been incurred.

9. Bradford L. Therrell, Jr., "Newborn Dried Bloodspot Screening," in *Newborn Screening Systems: The Complete Perspective*, ed. Carlie J. Driscoll and Bradley McPherson (San Diego, CA: Plural Publishing, 2010), 133–55, on 135. An estimated 15 to 20 percent of cases of CH are inherited. Genetics Home Reference, "Congenital Hypothyroidism," ghr.nlm.nih.gov/condition/congenital-hypothyroidism.

10. Jean H. Dussault and Claude Laberge, "Thyroxine (T4) Determination in Dried Blood by Radioimmunoassay: A Screening Method for Neonatal Hypothyroidism," *L'Union Médicale du Canada* 102 (1973): 2062–64; Jean H. Dussault, "The

Anecdotal History of Screening for Congenital Hypothyroidism," *Journal of Clinical Endocrinology and Metabolism* 84 (1999): 4332–34.

11. With the exception of cystinuria, all the other candidate conditions for screening were rarer—usually much rarer—than PKU.

12. Shackleton, "Report on Bill #3-126," 2.

13. Michael D. Garrick, Philip Dembure, and Robert Guthrie, "Sickle-Cell Anemia and Other Hemoglobinopathies: Procedures and Strategy for Screening Employing Spots of Blood on Filter Paper as Specimens," *New England Journal of Medicine* 288 (1973): 1265–68; Jane M. Benson and Bradford L. Therrell, Jr., "History and Current Status of Newborn Screening for Hemoglobinopathies," *Seminars in Perinatology* 34, no. 2 (2010): 134–44, on 136.

14. Committee for the Study of Inborn Errors of Metabolism, National Research Council, *Genetic Screening: Programs, Principles, and Research* (Washington, DC: National Academy of Sciences, 1975), table 6, p. 122. At the time, only Georgia, Kentucky, and Louisiana tested newborns for sickle-cell disease.

15. Benson and Therrell, "History and Current Status," 139.

16. D. S. Millington, N. Kodo, D. L. Norwood, and C. R. Roe, "Tandem Mass Spectrometry: A New Method for Acylcarnitine Profiling with Potential for Neonatal Screening for Inborn Errors of Metabolism," *Journal of Inherited Metabolic Disease* 13 (1990): 321–24.

17. Beth A. Tarini, Dimitri A. Christakis, and H. Gilbert Welch, "State Newborn Screening in the Tandem Mass Spectrometry Era: More Tests, More False-Positive Results," *Pediatrics* 118, no. 2 (2006): 448–56.

18. Michael S. Watson, Michele A. Lloyd-Puryear, Marie Y. Mann, et al., eds., "Towards a Uniform Screening Panel and System: Main Report," *Genetics in Medicine* 8, no. 5, suppl. (May 2006): 12S.

19. Bradford L. Therrell and John Adams, "Newborn Screening in North America," *Journal of Inherited Metabolic Disease* 30, no. 4 (2007): 447–65, on 447.

20. Committee for the Study of Inborn Errors, *Genetic Screening*, 104.

21. The report was commissioned by the Maternal and Child Health Bureau of the Health Resources and Services Administration of the US Department of Health and Human Services.

22. "High scoring" conditions, which were assigned to the core panel, were those receiving 1,200 to 1,799 points of a possible 2,100. The adoption of a multiplexing criterion and its weighting were justified on the grounds that "technology can add value to testing, particularly if it provides the ability to screen for many conditions in a single test. This can have public health importance above and beyond the features of the disease itself (i.e., by detecting secondary conditions) . . . Technologies with multiplexing capability offer improved efficiency and cost effectiveness to programs." Watson et al., "Towards a Uniform Screening Panel," 28–29S.

23. Ibid., 36–39S. The approaches are also known as "multiple-reaction monitoring" and "selective reaction monitoring."

24. Representative critiques include J. R. Botkin, E. W. Clayton, N. C. Fost, et al., "Newborn Screening Technology: Proceed with Caution," *Pediatrics* 17 (2006): 1793–99; President's Council on Bioethics, *The Changing Moral Focus of Newborn Screening: An Ethical Analysis by the President's Council on Bioethics* (Washington, DC: President's Council on Bioethics, 2008), bioethics.georgetown.edu/pcbe/reports/newborn_screening/index.html; Mary Ann Baily and Thomas H. Murray, "Ethics, Evidence, and Cost in Newborn Screening," *Hastings Center Report* 38, no. 3 (2008): 23–31.

25. Watson et al., "Towards a Uniform Screening Panel," 37S. For a somewhat different description of core screening criteria, see p. 43S.

26. Ibid., 13S, 43S.

27. US Department of Health and Human Services, "Secretary's Advisory Committee on Heritable Disorders in Newborns and Children," www.hrsa.gov/advisorycommittees/mchbadvisory/heritabledisorders. For current US statistics, see National Newborn Screening and Genetics Resource Center, "National Newborn Screening Status Report, Updated 07/11/12," genes-r-us.uthscsa.edu/sites/genes-r-us/files/nbsdisorders.pdf.

28. Centers for Disease Control and Prevention (CDC), "Hearing Loss in Children, Data and Statistics," www.cdc.gov/ncbddd/hearingloss/data.html. The value of newborn hearing screening is that early intervention dramatically improves outcomes for quality of speech, development of language, and long-term academic functioning. See also CDC, "Hearing Loss in Children, Historical Moments in Newborn Hearing Screening," www.cdc.gov/ncbddd/hearingloss/ehdi-history.html.

29. "Newborn Screening for Critical Congenital Heart Disease: Potential Roles of Birth Defects Surveillance Programs—United States, 2010–2011," *MMWR Morbidity and Mortality Weekly Report* 61, no. 42 (2012): 849–53, www.cdc.gov/mmwr/preview/mmwrhtml/mm6142a1.htm.

30. Jean Holt Koch, *Robert Guthrie: The PKU Story* (Pasadena, CA: Hope Publishing House, 1997), 93.

31. US Children's Bureau, US Department of Health, Education, and Welfare, *Proceedings of International Conference on Inborn Errors of Metabolism, May 30–June 3, 1966, Dubrovnik, Yugoslavia* (Washington, DC: US Department of Health, Education, and Welfare, 1967), participant list on 63–66, www.mchlibrary.info/history/chbu/19235.pdf.

32. Some of these programs were national, while others were organized at the regional, state, provincial, territorial, or even local level.

33. Therrell and Adams, "Newborn Screening," 453; see also table 2, p. 454.

34. Carmencita D. Padilla and Bradford L. Therrell, "Newborn Screening in the Asia Pacific Region," *Journal of Inherited Metabolic Disease* 30 (2007): 490–506, on 493–94.

35. Ibid., 495. See also D. Webster, "Newborn Screening in Australia and New Zealand," *Southeast Asian Journal of Tropical Medicine and Public Health* 34, suppl. 3 (2003): 69–70.

36. In Germany, blood-spot screening was introduced, with Guthrie's help, in Horst Bickel's screening laboratory in Marburg in 1961 and was implemented universally by 1968. Horst Bickel, "Phenylketonuria: Past, Present, Future," *Journal of Inherited Metabolic Disease* 3, no. 1 (1980): 123–32, on 127. In France, PKU testing began in 1966, with a national neonatal screening program established to coordinate coverage in 1978. J. L. Dhondt, J. P. Farriaux, M. L. Briard, et al., "Neonatal Screening in France," *Screening* 2, no. 2–3 (1993): 77–85. In Ireland, blood-spot screening began in the mid-1960s. Philip Mayne, "Pathology Department," in *The Children's Hospital, Temple Street: The Post Centenary Years (1972–2002)*, ed. Antoinette Walker (Dublin: Blackwater Press, 2002), 104–12. In the United Kingdom, where urine screening for PKU began in the 1950s, blood-based screening was adopted in 1969. M. Downing and R. Pollitt, "Newborn Bloodspot Screening in the UK—Past, Present and Future," *Annals of Clinical Biochemistry* 45 (2008): 11–17.

37. Katherine Bain and Clara Schiffer, *Experience with the Use of PL-480 Funds in Developing PKU Programs in Foreign Countries* (Washington, DC: Department of Health, Education, and Welfare, 1966), 7, 13.

38. Clara G. Schiffer, Program Analyst, Children's Bureau, to Donald Fraser, Hospital for Sick Children, Toronto, Dec. 2, 1964, USCB, NARA II, box 1000, file 4-5-11-5. Children's Bureau assistance in establishing programs in Poland (1964), Israel, Yugoslavia, and Pakistan is also mentioned in the memo "Proposed Form CB-80: Report on Phenylketonuria Blood Screening Program," Nov. 1, 1965, USCB, NCEMCH.

39. "This [Guthrie blood-spot testing] was the program that was exported to a number of other countries by means of several international meetings sponsored by the Children's Bureau with PL-480 funds in Poland, Egypt, Israel, Yugoslavia, and Germany. They included Poland, Egypt, Israel, Yugoslavia, Pakistan, China and Japan." Rudolf P. Hormuth, "Newborn Screening Systems" (paper presented at the 10th National Neonatal Screening Symposium, Seattle, WA, 1994), USCB, NCEMCH.

40. Therrell and Adams, "Newborn Screening," 456. For a current list of conditions screened in Canada, see Canadian Organization for Rare Disorders, "Newborn Screening in Canada Status Report," raredisorders.ca/documents/Canada NBSstatusupdatedMay52012.pdf.

41. On Germany, see Erik Harms and Bernhard Olgemöller, "Neonatal Screening for Metabolic and Endocrine Disorders," *Deutsches Ärzteblatt International* 108, no. 1–2 (2011): 11–22. Until 2005, only the tests for CH and galactosemia had been added by most German states. On the Netherlands, see Anne Marie C. Plass, "Extension of the Newborn Screening Programme in the Netherlands: Opinions of Prospective Parents, and Unintended Side Effects," in *Assessing Life: On the Organisation of Genetic Testing*, ed. Bernhard Wieser and Wilhelm Berger (Munich: Profil, 2010), 155–74, on 155. On Israel, see Shlomit Zuckerman, "The Expansion of Newborn Screening in Israel: Ethical and Social Dimensions" (PhD thesis, Case Western

Reserve University, 2009), 85–86, etd.ohiolink.edu/view.cgi/Zuckerman%20Shlomit.pdf?case1247156923. According to Zuckerman, the initiative for expansion came from two Americans with strong ties to Israel, and both the purchase of the tandem mass spectrometer and the training of personnel were funded by an American group. On Australia and New Zealand, see Bridget Wilcken and Veronica Wiley, "Newborn Screening," *Pathology* 40, no. 2 (2008): 104–15.

42. Health Service Executive (Ireland), "What Conditions Do We Test For?" www.hse.ie/eng/services/healthpromotion/newbornscreening/newbornbloodspotscreening/conditions.

43. For current statistics for the United Kingdom, see UK Newborn Screening Programme Centre, newbornbloodspot.screening.nhs.uk.

44. Peter Burgard, Martina Cornel, Francesco Di Filippo, et al., "Report on the Practices of Newborn Screening for Rare Disorders Implemented in Member States of the European Union, Candidate, Potential Candidate and EFTA Countries" (June 1, 2012), table 8.5, p. 53-7, ec.europa.eu/eahc/documents/news/Report_NBS_Current_Practices_20120108_FINAL.pdf.

45. J. Gerard Loeber, Peter Burgard, Martina C. Cornel, et al., "Newborn Screening Programmes in Europe: Arguments and Efforts Regarding Harmonization: Part 1. From Blood Spot to Screening Result," *Journal of Inherited Metabolic Disease* 35, no. 4 (2012): 603–11, on 605.

46. Ibid., table 2, pp. 608–9.

47. Jean-Louis Dhondt, "Expanded Newborn Screening: Social and Ethical Issues," *Journal of Inherited Metabolic Disease* 33, suppl. 2 (2010): S211–17, on S214.

48. J. M. G. Wilson and G. Jungner, *Principles and Practice of Screening for Disease* (Geneva: World Health Organization, 1968), whqlibdoc.who.int/php/WHO_PHP_34.pdf. James Maxwell Glover Wilson was principal medical officer at the Ministry of Health in London, and Gunnar Jungner was chief of Sahlgren Hospital's Clinical Chemistry Department in Gothenburg, Sweden. See also Anne Andermann, Ingeborg Blancquaert, Sylvie Beauchamp, and Véronique Déryc, "Revisiting Wilson and Jungner in the Genomic Age: A Review of Screening Criteria over the Past 40 Years," *Bulletin of the World Health Organization* 86, no. 4 (2008): 317–19.

49. Wilson and Jungner, *Principles and Practice*, 26.

50. Ibid., 27; see also 129.

51. National Institutes of Health, Eunice Kennedy Shriver National Institute of Child Health and Human Development, Office of Rare Diseases Research, and Office of Dietary Supplements, "Phenylketonuria Scientific Review Conference: State of the Science and Future Research Needs" (Bethesda, MD, Feb. 22–23, 2012).

52. Hans Zellweger and Alan Antonik. "Newborn Screening for Duchenne Muscular Dystrophy," *Pediatrics* 55, no. 1 (1975): 30–34.

53. NBS for Duchenne's was tried and eventually abandoned in Wales and Manitoba, though efforts to introduce it in the United States have been boosted by

recent advances in detection and in potential for treatment. See Jerry R. Mendell, Chris Shilling, Nancy D. Leslie, et al., "Evidence-Based Path to Newborn Screening for Duchenne Muscular Dystrophy," *Annals of Neurology* 71, no. 3 (2012): 304–13.

54. Jeffrey P. Brosco, Lee M. Sanders, Robin Dharia, et al., "The Lure of Treatment: Expanded Newborn Screening and the Curious Case of Histidinemia," *Pediatrics* 125, no. 3 (2010): 417–19. As noted in Bridget Wilcken, "Newborn Screening: How Are We Travelling, and Where Should We Be Going?" *Journal of Inherited Metabolic Disease* 34 (2011): 569–74, treatment in some jurisdictions continued long after it had been accepted elsewhere as not warranted, and few programs have actually been ended by lack of evidence of benefit (570).

55. David S. Millington, interview, Raleigh-Durham, NC, Dec. 17, 2008.

56. Wilcken, "Newborn Screening," 570.

57. Duane Alexander and Peter C. van Dyck, "A Vision of the Future of Newborn Screening," *Pediatrics* 117 (2006): S350–54, on S352.

58. Donald B. Bailey, Laura M. Beskow, Arlene M. Davis, and Deborah Skinner, "Changing Perspectives on the Benefits of Newborn Screening," *Mental Retardation and Developmental Disabilities Research Reviews* 12, no. 4 (2006): 270–79. Mara Buchbinder and Stefan Timmermans have recently described a different and completely unexpected type of family spillover effect from expanded NBS: the discovery of metabolic disease in the mother as a result of detection of an abnormal level of metabolites in the newborn. Mara Buchbinder and Stefan Timmermans, "Newborn Screening and Maternal Diagnosis: Rethinking Family Benefit," *Social Science and Medicine* 73, no. 7 (2011): 1014–18.

59. Scott D. Grosse, Coleen A. Boyle, Aileen Kenneson, et al., "From Public Health Emergency to Public Health Service: The Implications of Evolving Criteria for Newborn Screening Panels," *Pediatrics* 117, no. 3 (2006): 923–29.

60. Beth A Tarini, Dimitri A. Christakis, and H. Gilbert Welch, "State Newborn Screening in the Tandem Mass Spectrometry Era: More Tests, More False-Positive Results," *Pediatrics* 118 (2006): 448–56, on 452–53. Several recent works by sociologists examine the real-world experience of parents with expanded screening. See Rachel Grob, *Testing Baby: The Transformation of Newborn Screening, Parenting, and Policymaking* (New Brunswick, NJ: Rutgers University Press, 2011); Stefan Timmermans and Mara Buchbinder, *Saving Babies: The Consequences of Newborn Genetic Screening* (Chicago: University of Chicago Press, 2012).

61. Stefan Timmermans and Mara Buchbinder, "Patients-in-Waiting: Living between Sickness and Health in the Genomics Era," *Journal of Health and Social Behavior* 51, no. 4 (2010): 408–23. See also Timmermans and Buchbinder, *Saving Babies*.

62. Nathaniel Comfort, in *The Science of Human Perfection: How Genes Became the Heart of American Medicine* (New Haven, CT: Yale University Press, 2012), makes a similar point in relation to medicine in general, noting that the increasing emphasis on prevention and detection of "latent" disease is making everyone a patient (243).

63. Wilcken, "Newborn Screening," 570.

64. Burgard et al., "Report on Practices," 25.

65. Rodney J. Pollitt, in "International Perspectives on Newborn Screening," *Journal of Inherited Metabolic Disease* 29 (2006): 390–96, notes that the principles "are mainly formulated in qualitative terms with no clear end-points, which limits their applicability as a decision tool" (391), and for that reason, the UK national screening program has expanded the number of principles and made them more precise. For a description of current UK screening program appraisal criteria, see UK National Screening Committee, UK Screening Portal, "Programme Appraisal Criteria," www.screening.nhs.uk/criteria.

66. Jeffrey R. Botkin, "Assessing the New Criteria for Newborn Screening," *Health Matrix* 9 (2009): 163–86, on 165. See also Rodney J. Pollitt, "Introducing New Screens: Why Are We All Doing Different Things?" *Journal of Inherited Metabolic Disease* 30 (2007): 423–29.

67. Carla van El notes that in some cases, such as the Netherlands, the principles have also served to enable screening. Personal communication, Sept. 8, 2012.

68. Pollitt, "International Perspectives," 394.

69. Gerard Loeber notes that non-US companies that sell screening technology also try to influence policy and that in European countries where private screening laboratories exist, they "also try to charge parents extra fees for doing extra screening outside the official panel." Personal communication, Sept. 11, 2012.

70. The resolution and other associated materials are in the files of Neil R. M. Buist, Portland, OR. The resolution can be publicly accessed at Center for Media and Democracy, "Alec Exposed," alecexposed.org/w/images/1/1e/5B8-Resolution_to_End_State-Enabled_Newborn_Testing_Monopolies_Exposed.pdf. For a sympathetic perspective on Edwin Naylor's critique of "government-run labs," see "Medical Breakdown," Forbes.com, 2000, www.forbes.com/forbes/2000/0529/6513174a_print.html.

71. Forbes.com, "Medical Breakdown." For an account of NeoGen's legislative lobbying activities in Tennessee, see Ellen Wright Clayton, "Lessons to Be Learned from the Move toward Expanded Newborn Screening," in *Ethics and Newborn Genetic Screening: New Technologies, New Challenges*, ed. Mary Ann Baily and Thomas H. Murray (Baltimore: Johns Hopkins University Press, 2009), 125–35.

72. Michael Waldholz, "A Drop of Blood Saves One Baby; Another Falls Ill," Wall Street Journal Online, June 17, 2004, online.wsj.com/ad/article/philips/SB108741631056839034.html.

73. Steven Ertelt, "Critics Concerned New Economic Stimulus Bill Promotes Rationed Health Care," *LifeNews.com*, Feb. 10, 2009, www.lifenews.com/2009/02/10/nat-4830/?pr=1.

74. Typical of the rhetoric on conservative Christian and libertarian websites is the claim that the cost-effectiveness provision in the economic stimulus bill would "force the medical profession (under threat of sanctions) to allow the government

to decide who lives and who dies. Apparently, this is happening in Europe, Canada, Australia and now here." Christian Forums, "Homosexuality, Abortion, Contraception, Euthanasia, and Refusing Health Care" (Feb. 10, 2009), www.christianforums.com/t7340708.

75. Uwe E. Reinhardt, "'Cost-Effectiveness Analysis' and U.S. Health Care," *New York Times*, Economix blog, Mar. 13, 2009, economix.blogs.nytimes.com/2009/03/13/cost-effectiveness-analysis-and-us-health-care. See also "Evidence, Schmevidence," *Economist*, June 16, 2012, 32–34.

76. Burgard et al., "Report on Practices," 26.

77. Jennifer L. Howse and Michael Katz, "The Importance of Newborn Screening," *Pediatrics* 106 (2000): 595.

78. Mary Ann Baily, "Fair Distribution of Newborn Screening Costs and Benefits," in Baily and Murray, *Ethics and Newborn Genetic Screening*, 19–57, on 50.

79. Baily and Murray, "Ethics, Evidence, and Cost," 23.

80. Rachel Grob, "A House on Fire: Newborn Screening, Parents' Advocacy, and the Discourse of Urgency," in *Patients as Policy Actors*, ed. B. Hoffman, N. Tomes, M. Schlesinger, and R. Grob (New Brunswick, NJ: Rutgers University Press, 2011); see also Grob, *Testing Baby*, chap. 5.

81. In the European Union, Germany is apparently the only country to legislate the exact number of disorders to be screened for. Gerard Loeber, personal communication, Sept. 11, 2012.

82. For example, when newborn screening for PKU was introduced in the Netherlands in 1974, it was "without public discussion," as was also the case when the program expanded to include congenital hypothyroidism (1981) and congenital adrenal hyperplasia (2000). Carla van El, Toine Pieters, and Martina Cornel, "The Changing Focus of Screening Criteria in the Age of Genomics: A Brief History from the Netherlands," in *Assessing Life: On the Organisation of Genetic Testing*, ed. Bernhard Wieser and Wilhelm Berger (Munich: Profil, 2010), 203–24, on 206–7. Anne Marie C. Plass, in her essay in that volume, "Extension of the Newborn Screening Programme in the Netherlands: Opinions of Prospective Parents, and Unintended Side Effects" (155–74), notes that the expansion from 3 to 17 disorders in 2007 was also expert-driven (155).

83. See, for example, Mary A. Baily, "Fair Distribution of Newborn Screening Costs and Benefits," in Baily and Murray, *Ethics and Newborn Genetic Screening*, 19–57; Baily and Murray, "Ethics, Evidence, and Cost"; Rebecca Dresser, "Investigational Drugs and the Constitution," *Hastings Center Report* 36, no. 6 (2006): 9–10; Ezekiel J. Emanuel, "Drug Addiction: Cancer in the Courts," *New Republic*, July 3, 2006, 9–12.

84. Edwin W. Naylor, interview, Isle of Pines, SC, Oct. 23, 2008.

85. Philip Reilly, "Legal Issues in Genomic Medicine," *Nature Genetics* 7, no. 3 (2001): 268–71, on 269.

86. For example, "The technology could be expanded to screen for additional

disorders as mutational analysis or other multiplex technology become available, with decisions being based more on what not to screen for (perhaps Huntington disease) than on what to include." Alexander and van Dyck, "Vision," S353.

87. Stephen Cederbaum, "Newborn Screening: The Spigot Is Open and Threatens to Become a Flood," *Journal of Pediatrics* 151, no. 2 (2007): 108–10, on 109.

88. Committee for the Study of Inborn Errors, *Genetic Screening*, 88–93.

89. Neil A. Holtzman, "Anatomy of a Trial," *Pediatrics* 60 (1977): 932–34.

90. This paraphrases an argument in R. Rodney Howell, "We Need Expanded Newborn Screening," *Pediatrics* 117 (2006): 1800–1805, on 1800–1802.

Epilogue. "The Government Has Your Baby's DNA"

1. Archie Cochrane's *Effectiveness and Efficiency* (London: Nuffield Provincial Hospitals Trust) was published in 1972, and the term "evidence-based medicine" was first used in the 1990s.

2. In oral history interviews, Guthrie's politics were variously described as socialist, social-democratic, very liberal, liberal Democratic, and so forth. Although the specific characterizations differed, everyone who said they knew about his politics viewed them as left of center.

3. For example, Guthrie characterized the California program as "terribly fragmented and expensive" as the result of the health department's surrender to the private pathologists' lobby. He complained that, instead of the health department itself performing the test in a few centers, about a hundred private labs were doing it for a fee that then became part of the obstetrical bill. He commented, "I guess this demonstrates that a good law can be administered badly." Guthrie to Sen. William T. Conklin, NY State Senate, Oct. 12, 1973, Guthrie papers, box 37, folder "Letters (photocopies) re: Legislation 1973, 1976."

4. Herb Drill, "Buffalo's Jim Kelly Scores in Newborn Disease Effort," Herb's World Congress Report 2006, www.notaccessible.com/worldcongress/aab.html.

5. Committee for the Study of Inborn Errors of Metabolism, National Research Council, *Genetic Screening: Programs, Principles, and Research* (Washington, DC: National Academy of Sciences, 1975), 92.

6. E. R. McCabe, S. Z. Huang, W. K. Seltzer, and M. L. Law, "DNA Microextraction from Dried Blood Spots on Filter Paper Blotters: Potential Applications to Newborn Screening," *Human Genetics* 75 (1987): 213–16.

7. Bradford L. Therrell, Jr., and W. Harry Hannon, "Newborn Dried Blood Spot Screening: Residual Specimen Storage Issues," *Pediatrics* 129, no. 2 (2012): 365–66.

8. For a recent analysis of the situation in the United States, see Michelle H. Lewis, Aaron Goldenberg, Rebecca Anderson, et al., "State Laws Regarding the Retention and Use of Residual Newborn Screening Blood Samples," *Pediatrics* 127, no. 4 (2011): 703–12. The NIH has funded the Newborn Screening Translational Research Network (www.nbstrn.org) to develop model policies and facilitate access for researchers.

9. Neil A. Holtzman, "Public Participation in Genetic Policymaking: The Maryland Commission on Hereditary Disorders," in *Genetics and the Law II*, ed. A. Milunksy and G. J. Annas (New York: Plenum, 1980), 247–55, on 253.

10. Ruth A. Faden, Judith Chwalow, Neil A. Holtzman, and Susan D. Horn, "A Survey to Evaluate Parental Consent as Public Policy for Neonatal Screening," *American Journal of Public Health* 72 (1982): 1347–51, on 1349.

11. Ruth A. Faden, Neil A. Holtzman, and A. Judith Chwalow, "Parental Rights, Child Welfare, and Public Health: The Case of PKU Screening," *American Journal of Public Health* 72 (1982): 1396–1400, on 1397.

12. For a good overview of the ethics of informed consent for NBS, see Lainie F. Ross, "Mandatory versus Voluntary Consent for Newborn Screening?" *Kennedy Institute of Ethics Journal* 20, no. 4 (2010): 299–328.

13. Elizabeth Cohen, "The Government Has Your Baby's DNA," *CNN Health*, Feb. 4, 2010, www.cnn.com/2010/HEALTH/02/04/baby.dna.government/index.html

14. Dana Barnes, "Texas DNA Showdown," *Mayborn*, Aug. 12, 2010, www.themayborn.com/article/texas-dna-showdown.

15. J. Gerard Loeber, "Storage, Management, and Use of DBS in The Netherlands" (paper presented at Workshop on Storage and Secondary Use of Residual Dried Bloodspots, VU University Medical Center, Amsterdam, Apr. 10, 2012). There were also many other newspaper articles, TV discussions, and questions put to the Minister of Health in Parliament. Gerard Loeber, personal communication, Sept. 11, 2012.

16. Jennifer Couzin-Frankel, "Science Gold Mine, Ethical Minefield," *Science* 324 (2009): 166–68, on 167.

17. That ruling was upheld in 2011 by the Department of Health. See Health Service Executive (Ireland), "Information on Stored Newborn Screening Cards," www.hse.ie/eng/services/healthpromotion/newbornscreening/newbornbloodspotscreening/nscs.html.

18. Mary Carmichael, "A Spot of Trouble," *Nature* 475 (2011): 156–58, on 157.

19. Twila Brase, "Newborn Genetic Screening: The New Eugenics?" Citizens' Council on Health Care Report (Apr. 2009), www.cchfreedom.org/pdf/NBS_EUGENICS_REPORT_Apr2009_FINAL.pdf.

20. The organization is Verdedigt Digitale Burgerrechten ("Bits of Freedom"), www.bof.nl/home/english-bits-of-freedom.

21. The case can be found at caselaw.findlaw.com/mn-supreme-court/1585739.html.

22. Meredith Wadman, "Minnesota Starts to Destroy Stored Blood Spots," *Nature*, Feb. 3, 2012, www.nature.com/news/minnesota-starts-to-destroy-stored-blood-spots–1.9971.

23. Carmichael, "Spot of Trouble," 157–58.

24. Texas Civil Rights Project, www.texascivilrightsproject.org/?p=3170.

25. A second lawsuit was filed when journalists discovered, from records released

after the settlement, that the state had given eight hundred anonymized samples to the Armed Forces DNA Identification Laboratory to help create a national mitochondrial DNA database. Emily Ramshaw, "DSHS Turned over Hundreds of DNA Samples to Feds," *Texas Tribune*, Feb. 22, 1010, www.texastribune.org/texas-state-agencies/department-of-state-health-services/dshs-turned-over-hundreds-of-dna-samples-to-feds.

26. The "Newborn Screening Saves Lives Act" is also the title of legislation that strengthened the federal role in screening policy and established grant programs for education, outreach, coordination of follow-up care, and research, especially on candidate conditions for screening.

27. American College of Medical Genetics, "Position Statement on Importance of Residual Newborn Screening Dried Blood Spots" (approved Apr. 29, 2009), www.acmg.net/StaticContent/NewsReleases/Blood_Spot_Position_Statement2009.pdf.

28. Preserving the Future of Newborn Screening, www.newbornbloodspots.org.

INDEX

Page numbers followed by f indicate figures and those followed by t indicate tables.

Alexander, Duane, 64, 194
AMA (American Medical Association), 63, 233n17; opposition to mandatory testing, xix, 84–85, 90, 246n46, 247n53
American Academy of Pediatrics (AAP): Committee on the Handicapped Child, 72–73, 243n15; Committee on Nutrition, 86; and expanded NBS, 186; and informed consent, 207; opposition to mandatory PKU screening, xix, 72, 89, 181, 242n2
American Association on Mental Deficiency (later American Association on Intellectual and Developmental Disabilities), 24
American College of Medical Genetics (ACMG—later American College of Medical Genetics and Genomics), 185–86, 190–91, 196
American Legislative Exchange Council (ALEC), 197
Amino Acid Composition of Proteins and Foods (Block and Bolling), 16
Angel Unaware (Rogers), 33
Armstrong, Marvin, 42, 234n20
Asilomar conference, 95
aspartame, 114–15, 254n13
Association for the Aid of Crippled Children, 52, 61
Asylums (Goffman), 171

bacterial inhibition assay (BIA). *See* Guthrie test
Bank-Mikkelsen, Niels Erik, 171–72

Bashford, Alison, 157
Bateson, William, 227n4
Beleno et al. v. Texas Department of State Health Services et al. (2009), 208
Bell Curve, The (Herrnstein and Murray), 104
Bessman, Samuel P.: criticism of mandatory PKU screening, 73, 79, 82, 90, 181, 247n53, 272n5; influence of, 85–87; justification hypothesis, 81–82; and maternal PKU, 145–46; scientific and technical contributions (non-PKU), 85, 247n52; unsuccessful treatment of child with tyrosine, 81, 245n37
BH_4. *See* sapropterin dihydrochloride
Bickel, Horst, 39–41, 233n12
Binet, Alfred, 27
Bittersweet (Feudtner), xvi, 253n46
Block, Richard J., 16, 36, 38, 233n8
Block and Bolling method (for estimating the amino acid content of proteins), 16, 38
Boggs, Elizabeth M., 51–53, 61, 86–87, 156, 238n21, 239n24
Bolling, Diana, 16, 38
Botkin, Jeffrey, 195
Boulding, Kenneth, 163
Brase, Twila, 209–10
Brearder v. State of Minnesota (2010), 209
Britain. *See* United Kingdom
Buchbinder, Mara, 195, 277n58
Buck, Carol, 1–4, 3f, 7, 10, 21, 23, 43, 225n8, 226n16
Buck, Janice, 2, 3f

Buck, John Lossing, 2, 3f, 7, 225n4, 226n16
Buck, Pearl S. (née Sydenstricker), 1–3, 3f, 5–7, 9, 21, 39, 60, 225n4, 226n16; "The Child Who Never Grew" (essay), 1, 5; *The Child Who Never Grew* (book), 2, 6–7, 32–33, 225n4, 225n8; and Vineland Training School, 2–3, 4f, 20–21
Buist, Neil R. M., 118–19, 129, 263n34
Burt, Cyril, 102

Centerwall, Willard R., 43–44, 144, 225n8, 241n55
Childs, Barton, 96
"Child Who Never Grew, The" (Buck, Pearl S.), 1, 5
Child Who Never Grew, The (Buck, Pearl S.), 2, 6–7, 32–33, 225n4
Citizens' Council on Health Care (later Citizens' Council for Health Freedom), 209–10
Clarke, Angus, 107
cognitive impairment: association with PKU, 4–5, 13, 15–16, 110; definitions of, 27–28; and disability rights movement, 171–73; and emergence of parent groups, 30–33; and intellectual disability, xx–xxi; and intelligence tests, 27, 30; physician disinterest in, 63; as social problem, 25–27; and social stigma, 3, 5–7, 30, 32–33; and training schools, 23–25, 27–28. *See also* eugenics; parent/patient advocacy organizations
Collaborative Study of Children Treated for Phenylketonuria, 70–80, 82–83
Collins, Francis, 107
Comfort, Nathaniel, 18
Committee for the Study of Inborn Errors of Metabolism (National Academy of Sciences), 88, 205, 207
congenital hypothyroidism (CH), 183–84, 190–91, 272n9
Cooke, Robert E., 59–61, 64–65, 156, 237n15, 239n28; *The Dark Corner*, 59–60, 238n20; relationship with Eunice and Sargent Shriver, 59–61, 238n20
Cooper, Joseph D., 87–88
Cushing, Richard Cardinal, 59

Cymogram (Allen & Hanbury), 43, 117–18
cystic fibrosis, 93, 191, 249n5

Dark Corner, The (Cooke), 59–60, 238n20
Dent, Charles, 42, 141–42, 233n20
Dhondt, Jean-Louis, 191
diabetes, 60, 134, 165, 191, 253n46, 256n39; compared with PKU, xvi, 110, 126–27
disability rights movement, 170–73, 178, 269n50
disease-specific advocacy groups (contemporary), 186, 197, 200, 205, 210
District of Columbia. *See* Washington, DC
Dix, Dorothea, 24
Dobzhansky, Theodosius, 160
Doll, Margaret (niece of Robert Guthrie), 49
Douglas, Mary, 132–33
Duchenne muscular dystrophy, 192, 196, 203
Dussault, Jean, 183
Dybwad, Gunnar, 51, 87, 265n2

Efron, Mary L., 77, 83–84, 246n44, 246n48
Egan, Mary, 146
Egeland family, 10–13, 39
Ehrlich, Paul, 163
Eliot, Martha, 58
Erikson, Erik and Joan, 32
Esquirol, Jean-Étienne Dominique, 26–27
Ethics of Genetic Control, The (Fletcher), 161–62
EU (European Union), 191, 195, 198, 279n81
eugenics: and abortion for fetal anomalies, 162, 172–77, 269n52; and bioethics, 175–76; and carriers, xviii–xix, 7, 16–20, 156–58, 164–69; changes in meaning, xxiv, 19–20; and controversy over storage and use of Guthrie cards, 209; and disability rights movement, 171–73; and feminism, 173–74; and maternal PKU, 146–48; and medical progress, 164–69; mid-1940s–1950s, 17–19; 1960s–1980s, 94–96, 156–70; and Lionel Penrose, 17–19; PKU screening as break with, xviii–xix, 156; and population control, 162–64; prewar, 6–7, 28–29; and reproductive genetic services, 174–75, 269n52

Index

Fanconi, Guido, 39
FDA (US Food and Drug Administration), 43, 114, 135
ferric chloride test: adapted by Centerwall, 43–44, 234n36; excitement surrounding, xviii, 44–46, 80–81; Phenistix modification of, 44; problems associated with, 46, 49; recommended use in population-wide screening, 41–42, 45–46, 68; use by Bickel, 39; use by Følling, 12, 43, 227n6; use in United Kingdom, 8, 41–45; use in United States, 8, 42–45, 52
Feudtner, Chris, xvi, 253n46
Fisch, Robert O., 261n13, 262n32
Fletcher, Joseph, 161–62
Fogarty, John, 58
Følling, [Ivar] Asbjørn: and carrier detection, 15, 165; discovery of PKU, 10–15, 43, 227n6; and maternal PKU, 141, 155; receives Joseph P. Kennedy, Jr. Foundation Award, 63, 64f, 65f; reception of discovery, 20–21; reputation today, 20, 68; training and early career, 11, 12f, 226n2
"Food for Peace" (US Agricultural Trade and Development Assistance Act of 1954; PL-480), 188–90, 275n39
Fost, Norman, 243n15

galactosemia, 46, 164, 180, 182, 184, 188, 268n31, 275n41
Garrick, Michael, 183. *See also* sickle-cell disease
Garrod, Archibald, 11–12, 14, 227n4
genetic counseling, 161, 174–78
"genetic exceptionalism," 100
Genetic Information Nondiscrimination Act (2008), 100
Genetic Screening: Programs, Principles, and Research (1975), 88, 97
Gerrard, John, 37, 39–41, 233n12
Gilbert, Walter, 105
glycomacropeptide (GMP), 119
Goffman, Ervin, 171
Griffiths, Ruth, 141
Grob, Rachel, 199

Grosse, Scott, 194
Guthrie, John, 46–47
Guthrie, Robert, xviii, 35, 63, 66, 74–76, 88, 199; on California NBS program, 272n8, 280n3; cognitive impairment in family, 46, 49; develops bacterial inhibition assay for PKU, 46–53, 48f, 50f, 235n41; develops other bacterial assays, 179–82; international travels, 187–88, 189f; and maternal PKU, 142; and NARC, 47, 51–53, 68, 265n2; as outsider, 75; political orientation, 205, 280n2; rejection of eugenics, xviii, 156; responses to critics, 88–91, 248n63; on state laboratories, 70–71
Guthrie cards, 50f, 51; controversies over storage and use, 205–11; disparities in policy and practice, 206; and informed consent, xv, 206–9; research value of, 88, 99, 203, 205–8, 251n20. *See also* Phillips Punch Index Machine
Guthrie test (bacterial inhibition assay [BIA]), 44, 48f, 50f, 271n1; advantages over ferric chloride test, 49, 51; Children's Bureau field trial, 51–53, 75; development of, 47–54; issues of test reliability and validity, xix, 75–76; patenting/licensing of, 52, 205; poster campaign for, 52f, 65f, 89; Ada Susi role, 236n52; viewed as breakthrough, 63–66

Hanley, William B., 77–78
Hansen, Holger, 143, 145, 147, 155, 262n24
Hardin, Garrett, 164
Harding, Cary, 113, 127
Henderson, Lawrence J., 11
Herrnstein, Richard J., 101, 103–4
Hickmans, Evelyn, 40
histidinemia, 180, 193, 227n54
homocystinuria, 66, 180, 182, 240n42
Hormuth, Rudolph P., 181
Horner, Frederick, 42
Hopkins, Sir Frederick Gowland, 15, 21
Howe, Samuel G., 24, 26
Howell, R. Rodney, 186
Hsia, David Yi-Yung, 86

Human Genome Project: controversies surrounding, 104–6; legitimated by case of PKU, 106–9

informed consent, 204; for NBS, xv, 97–100, 202–3, 206–9; for storage and use of Guthrie cards, xv, 99–100, 203, 208–11
Ireland, 188, 190–91, 208–9, 275n36

Jaenisch, Rudolf, 95
Jensen, Arthur, 101, 103–4
Jervis, George A.: coins expression "phenylpyruvic oligophrenia," 232n4; and diet therapy, 36; interest in carrier detection, 165; research on biochemistry and genetics of PKU, 14–17
Jones, Mary, 39–41
Jones, Sheila, 39–40
Jonsen, Albert, 164–65, 176
Joseph P. Kennedy, Jr. Foundation, 59, 63, 64f, 65f

Kallikak Family, The (Goddard), 6, 28–29
Kamin, Leon, 102
Kang, Ellen, 78, 248n63
Kaufman, Seymour, 224n7
Kennedy, John Fitzgerald, 58, 60–61, 62f, 64f, 68
Kennedy, Joseph, Sr., 58–59
Kennedy, Joseph P., Jr., 59–60
Kennedy, Rose, 59–60
Kennedy, Rosemary, 58, 60, 238nn21–22
Kennedy administration, 58–63
Kennedy-Prouty bill (1965), 86–87
Kirkman, Henry N., 106, 150–51, 154–55
Koch, Jean Holt, 40
Koch, Richard K., 146
Kuvan (sapropterin dihydrochloride), 108–9, 155, 252n45

Lederberg, Joshua, 159
Lenke, Roger, 149–51
Levy, Harvey L., 149–51, 262n24
Lindee, M. Susan, 158
Lofenalac (Mead Johnson), 43, 52f, 77, 117, 144
Low Protein Food List for PKU (Schuett), 115

Mabry, C. Charlton, 86, 142
MacCready, Robert A., 66–67, 88, 90, 156
MacDonald, Anita, 125–26
March of Dimes (formerly National Foundation for Infantile Paralysis): and expanded NBS, 186, 198; 1930s–1960s, 31–32, 86
Masland, Richard L., 58, 61, 63, 237n17
maternal PKU (MPKU), 140–55; adherence to diet, 144, 154–55; cause of harm in, 141; challenges of diet during pregnancy, 144, 152–54, 154t; discussion of in 1960s and 1970s, 141–46, 262n24; and family planning, 146–48, 157, 177, 259n2, 264n53; Kirkman projection of "rebound effect" (1982), 150–51, 154; Lenke and Levy survey (1980), 149–50; outcomes in, 149, 151–52, 154–55, 263n40; uncertainty about diet safety and efficacy, 144–46
Maternal PKU Collaborative Study (MPKUCS), 148–49, 151–52, 154, 264n48
Mayo, Leonard, 61
McCabe, Edward R. B., 205
MCADD (medium-chain acyl-CoA dehydrogenase deficiency), 191
McGrath sisters (Kammy and Sheila), 52f, 62f, 65f, 89
Medawar, Peter B., 166–68, 268nn33–34
mental retardation (MR). *See* cognitive impairment
Mental Subnormality (Masland), 61
Millington, David, 184, 193
Mol, Annemarie, 134
Montagu, Ashley, 160–61
Montoya, Joseph, 69
mortality transition, 57
MPKU. *See* maternal PKU
MS-MS. *See* tandem mass spectrometry
Muller, H. J. (Hermann Joseph), 158–60, 168
Murray, Charles, 104

National Association for Retarded Children (NARC—originally National Association for Retarded Children; later National Association for Retarded Citizens, The ARC), 59, 66, 81, 237n15,

238n21; and Children's Bureau field trials of Guthrie test, 51–52; concern about access to treatment, 87–88, 205; founding and growth, 31–34, 57; and Robert Guthrie, 47, 51–53, 68, 70, 265n2; internal tensions, 265n2; name changes, 31; political activities, xix, 53, 57–58, 60, 65f, 67–68, 89–90, 199, 201–2, 239n36, 240n43; and President's Panel on Mental Retardation (PPMR), 61–62; receives Joseph P. Kennedy, Jr. Foundation award, 63, 64f, rejects eugenics, xviii, 156; and urine testing, 45–46
National Institute of Child Health and Human Development (NICHD), 59, 64–65, 151
National Institutes of Health. *See* NIH
nature-nurture debate, xiv–xv, 93, 101–4, 109, 251n29
Naylor, Edwin W., 180, 196–97, 236n52
NBS (newborn screening): anomaly in United States, xiv, 54, 187; for congenital heart disease, 186–87; for hearing impairment, 186–87, 274n28; physician attitudes toward, 84, 207, 247n49; separation of laboratory screening from clinical care in United States, 90–91
NBS, international: Australia, 9, 187–88, 190, 193; Canada, 9, 188; Cardiff, Wales, 45; China, 190; continental Europe, 9, 188, 190–91; France, 190–91, 275n36; Germany, 190, 275n36; Ireland, 190–91, 208–9, 275n36; Israel, 188, 190, 276n41; Japan, 188, 190; New Zealand, 9, 188–89; nonindustrialized countries, 188–90; Quebec, 183, 249n68, 271n1; Scandinavia, 191; Spain, xviii, 191; United Kingdom, xvii–xviii, 8–9, 42, 44–46, 190–91, 196, 199, 275n36
NBS expansion, US (1970s and 1980s), 179–83, 185; and congenital hypothyroidism, 183; constraints on, 184; and cost-effectiveness criteria, 88, 189; and cost-efficiency, 181–82; and Phillips Punch Index Machine, 180–81; and sickle-cell disease, 183
NBS expansion, US (1990s–): and American College of Medical Genetics, 185–86, 190, 196; arguments in favor of, 185–86, 193–94, 201–2; and commercial interests, 196–97, 204–5; and cost-effectiveness criteria, 197–98, 278n74; critiques of, 198–99, 201–2; disparities accompanying, 184–85; endorsements of, 186–87; and "patients-in-waiting," 195; and tandem mass spectrometry, 184–85, 190, 193, 195–96; and "urgency narrative," 199; and Wilson and Jungner screening principles, 191–96, 211
NBS technologies: chromatography, 51, 83, 180, 271n1; fluorometry, 51, 180, 271n1; tandem mass spectrometry (MS-MS), 51, 184–85, 190, 193, 195–97, 200; whole genome sequencing, 195
newborn screening. *See* NBS
NIH (National Institutes of Health), 59, 61, 65, 112, 237n17, 280n8
Nixon, Richard, 96

O'Conner, Basil, 31

parent/patient advocacy organizations: before 1950, 30–31; 1950s–1970s, 8, 31–34, 57–58, 63, 90; 1990s–, 186, 197–98, 200, 205, 210. *See also* National Association for Retarded Children
Parent Teacher Association (PTA), 32, 68
Pass, Kenneth, 66
"patients-in-waiting" (Timmermans and Buchbinder), 195
Pauling, Linus, x, 159, 166–68
Pemberton, Stephen, 93
Penrose, Lionel S., 92, 150; *The Biology of Mental Defect*, 36–37, 232n3; and Carol Buck, 20–21; concern about carriers, xviii–xix, 14, 17, 19, 165, 229n32; critique of eugenics, xviii, 15, 17–19, 29, 36, 92; dietary experiments, 15–16, 38, 228n22; "Phenylketonuria: A Problem in Eugenics," 18–20, 36; research on phenylketonuria, 14–15, 17; scepticism about dietary treatment, 15–16, 36–37, 40–41, 232nn3–4; and terminology, 14–15

Perry, Thomas, 147, 262n31
phenylalanine hydroxylase (PAH) enzyme, 5, 15, 74, 223n7; and sapropterin, 109, 252n45
Phillips Punch Index Machine, 180–81, 182f, 271n3
PKU (phenyloketuria): clinical and genetic heterogeneity, 5, 108, 225n10; forms of, 5, 74–75; life expectancy, 165; pathogenesis, 4–5, 13–16, 223n7; symptoms of untreated disease, 4, 10, 39–40
PKU diet: and aspartame, 114–15, 254n13; "diet for life," 111–12; experiments, 1930s–1950s, 15–16, 35–42, 228n22; "free foods," 115; and group identity, 130–33; guidelines, 111–14; improvements in, 112, 118–22; need for monitoring, 112–13; need for planning, 127, 133–34, 138–39; organoleptic properties of, 112, 118–19; and taste preferences, 128–30, 130t, 131t
—adherence, 256n35; in maternal PKU, 144; in PKU, 110, 113, 125–27, 135
—challenges: for adolescents and adults, 125–39; for children, 122–25, 125t; for pregnant women, 144, 152–54
—components: low-protein foods, 122, 123t; protein restriction, 113–17; protein substitute / "medical food" / "formula," 117–22
—controversies, 1960s and 1970s: over duration, 79–80; over efficacy, 80–83; over possible harms of, 76–79; over who should be treated, 74–75
—cost: as barrier to treatment, 135–38; and FDA reclassification of medical foods, 135; insurance coverage of, 136–37; low-protein foods, 123t; prior to availability of commercial formulas, 15, 38
—outcomes: in maternal PKU, 149, 151–52, 154–55, 263n40; in PKU, 109, 127–29, 257n40, 258n51
PKU incidence/prevalence, 18–19, 43, 53, 74, 189, 223n1, 272n7
PKU mandatory testing: advocacy of, 64–69, 88–90; opposition from organized medicine, 8–9, 83–84, 90, 246n46; scepticism of, 8–9, 72–90, 97, 101, 109
PL-480 program. *See* "Food for Peace"
polio, 31–32, 55–56, 60, 62, 92–93
Pollitt, Rodney, 196, 278n65
Population Bomb, The (Ehrlich), 163
Preserving the Future of Newborn Screening Coalition, 210–11
President's Panel on Mental Retardation (PPMR), 61–63, 62f

Quastel, Juda, 14, 228n22
Quebec, province of, 183, 249n68, 271n1

Ramsden, Edward, 169
Ramsey, Paul, 161–62, 170
Reed, Sheldon, 161
Reilly, Philip, 200
reproductive counseling. *See* genetic counseling
residual dried blood spots (DBS). *See* Guthrie cards
Rockefeller Foundation, 21
Roe v. Wade (1973), 95, 147, 174
Rogers, Dale Evans, 32–33; *Angel Unaware*, 33
Roosevelt, Franklin Delano, 31

sapropterin dihydrochloride (tetrahydrobiopterin, BH_4, or Kuvan [trade name]), 108–9, 155, 252n45
Scriver, Charles R.: critique of eugenics, 160; and PKU as paradigm, 9, 92; and Quebec Network of Genetic Medicine, 249n68
Secretary's Advisory Committee on Heritable Disorders in Newborns and Children (SACHDNC), 186–87, 197
Séguin, Édouard, 23–24
Shriver, Eunice Kennedy: "Hope for Retarded Children" (1962), 60; influence on federal policy, 59–61, 64–65, 239n28; relationship with Robert E. Cooke, 59–60
Shriver, Sargent, 59, 239n28
sickle-cell disease (SCD): controversies

in 1970s, 96, 98, 100, 207, 250n11; National Sickle Cell Anemia, Cooley's Anemia, Tay-Sachs, and Genetic Diseases Act (1976), 96; National Sickle-Cell Anemia Control Act (1972), 96; NBS for, 183–84, 191, 273n14; and Linus Pauling, 81, 167–68; penicillin prophylaxis for, 183; screening for disease and carrier status, 96–97
Smith, Isabel, 80
Streamer, Charles, 42
Susi, Ada, 47, 53, 236n52
syphilis, 56, 176

tandem mass spectrometry (MS-MS), 51, 184–85, 190, 193, 195–97, 200
Tarjan, George, 61
Tay-Sachs disease, 20, 96–97, 161
tetrahydrobiopterin (BH$_4$). *See* sapropterin dihydrochloride
Texas Civil Rights Project, 210
Therrell, Bradford L., Jr., 180–81
Timmermans, Stefan, 195, 277n58
Trudeau, Arthur, 57–58
Tyler, Frank, 42
tyrosinemia (hereditary), 82, 180, 183, 188

United Kingdom, 9, 19, 171, 174, 275n36; and Cymogram (formula), 43, 117–18; dietary experiments, 1930s–1940s, 15–16, 35–42; dietary guidelines, 80, 115; diet practices of, vs. United States, 117–18; home health visits, 44, 46; NBS expansion, 190–91; pioneers NBS (using urine), xvii–xviii, 8, 44–46; screening policy/practice of, vs. United States, 46, 196, 199
United States Children's Bureau (UCSB), 57–58, 67, 69–70, 72, 78, 82, 86, 141, 181; field trial of Guthrie test, 51–53, 66, 180, 272n7; and maternal PKU, 141–42, 146; and PL-480 ("Food for Peace") program, 188–90, 275n39; and urine screening, 45; and PKUCS, 82–83
United States Collaborative Study of Children Treated for Phenylketonuria (PKUCS), 79–80, 82–83
urine test. *See* ferric chloride test

Van Slyke, Donald, 11
Vineland Training School, 6, 23–24, 26; and Carol Buck, 1–4, 4f, 21, 225n8; and Pearl Buck, 2–3
Vulliamy, David, 37

Wailoo, Keith, 93
Waisbren, Susan E., 219
Waisman, Henry, 145
Warner, Robert, 47, 236n52
Washington, DC: reinstates screening program, 183; repeals screening law, 181
Waterston, Robert, 106–7
Watson, James D., 105
wet-diaper/nappy test. *See* ferric chloride test
Wilcken, Bridget, 193, 195
Wilson, Edward O., 103
Wilson and Jungner screening principles (1968), 191–92, 211; critiques of, 193–94; defenses of, 194; disparate interpretations of, 196–97; and PKU, 192; status internationally, 196
Wolff, Otto, 80
Woo, Savio, 151, 156
Woolf, Louis I.: consulted by Horst Bickel, 39–40; invents method to remove phenylalanine from casein, 37–39; on Lionel Penrose, 37; treats newborn for PKU, 41–42; on US-UK differences in diet, 117–18
World Health Organization (WHO), 126, 147, 192